T0329554

Emerging Computing Paradigms

Emerging Computing Paradigms

Principles, Advances and Applications

Edited by

Umang Singh
ITS, Ghaziabad (U.P.), India (deceased)

San Murugesan
BRITE Professional Services, Sydney, Australia

Ashish Seth
Inha University, Incheon, South Korea

This edition first published 2022
© 2022 John Wiley & Sons Ltd

The right of Umang Singh, San Murugesan, and Ashish Seth to be identified as the authors of this has been asserted in accordance with law.

Registered Offices
John Wiley & Sons, Inc., 111 River Street, Hoboken, NJ 07030, USA
John Wiley & Sons Ltd, The Atrium, Southern Gate, Chichester, West Sussex, PO19 8SQ, UK

Editorial Office
The Atrium, Southern Gate, Chichester, West Sussex, PO19 8SQ, UK

For details of our global editorial offices, customer services, and more information about Wiley products visit us at www.wiley.com.

Wiley also publishes its books in a variety of electronic formats and by print-on-demand. Some content that appears in standard print versions of this book may not be available in other formats.

Library of Congress Cataloging-in-Publication Data
Names: Singh, Umang, 1982-2021, editor. | Murugesan, San, 1978- editor. |
    Seth, Ashish, editor.
Title: Emerging computing paradigms : principles, advances and applications /
    edited by Umang Singh, ITS, Ghaziabad (U.P.), India (deceased), San
    Murugesan, BRITE Professional Services, Ashish Seth, Inha University,
    Incheon, South Korea.
Description: First edition. | Hoboken : John Wiley & Sons, 2022. | Includes
    bibliographical references and index.
Identifiers: LCCN 2021060411 (print) | LCCN 2021060412 (ebook) |
    ISBN 9781119813408 (hardback) | ISBN 9781119813415 (pdf) |
    ISBN 9781119813422 (epub) | ISBN 9781119813439 (ebook)
Subjects: LCSH: Computer science--Technological innovations. |
    Computer networks--Technological innovations.
Classification: LCC QA76.24 .E43 2022 (print) | LCC QA76.24 (ebook) |
    DDC 004--dc23/eng/20220201
LC record available at https://lccn.loc.gov/2021060411
LC ebook record available at https://lccn.loc.gov/2021060412

Cover image: © Rachael Arnott/Shutterstock
Cover design by Wiley

Set in 9.5/12.5pt STIXTwoText by Integra Software Services, Pondicherry, India.

C9781119813408_040722

Dedication

We dedicate this book to all the **computing pioneers,** *past and present, who laid foundation for modern computing and contributed to advances in all aspects of computing over the past 75 years—the immense benefits of which not only humans but also "things" around us enjoy today and will continue to relish in the future.*

Next, we dedicate this volume to **Umang Singh,** *our beloved friend and co-editor of this book, who was a key player in bringing out this book but is no longer with us to see it in print. She couldn't win the battle against COVID-19 and succumbed to it following an apparent victory that sadly didn't last long. Her memories and contributions to computing as an academic, as a researcher, and as an excellent human being, will live with us—readers of this book and the community at large.*

Contents

Preface

2021 marks the 75th anniversary of the first general purpose electronic digital computer, the 50th anniversary of the microprocessor, and the 40th anniversary of the IBM Personal Computer. These historic developments formed the foundation for amazing continuing advances in computing and IT. In a span of 75 years, from an unproven technology to one that is embedded deeply into every aspect of our work and our daily lives, computers have advanced significantly.

The history of the modern computer has its origin to ENIAC (Electronic Numerical Integrator and Computer), the first general purpose programmable electronic (vacuum-tube) computer, unveiled to public on February 14, 1946. John Mauchly and J. Presper Eckert at the University of Pennsylvania developed it secretly for the US Army to calculate ballistic trajectory tables more effectively than the mechanical differential analyzers in use at the time. Besides using it for ballistics trajectory research, ENIAC was also used for Monte Carlo simulations, weather predictions, and early hydrogen bomb research. On the eve of unveiling of ENIAC, the US War Department put out a press release hailing it as "a new machine that is expected to revolutionize the mathematics of engineering and change many of our industrial design methods." Without doubt, electronic digital computers did transform irrevocably engineering and mathematics, and also every other conceivable domain.

In 1958, Jack Kilby of Texas Instruments invented the first integrated circuit (IC). Adopting large-scale-integration, very large-scale-integration, and ultra-large-scale integration, the density of ICs continued to increase, closely following the Moore's Law—the number of transistors in a chip will approximately double every 24 months. In 1971, Intel Corporation, released the world's first microprocessor, a complete general-purpose central processor unit of a computer on a single IC. Amazing developments in hardware, software, and networking and communication followed, and changed the world irrevocably.

As computers have evolved to redefine and transform almost every area of our lives in the past 75 years, they still function on the same fundamental computational concepts envisaged at the beginning. As demands on computing, storage, and communication continue to escalate, digital computers based on silicon and conventional architecture approach their limits and face issues related to economics and reliability. Thus, certain kinds of problems in domains such as weather forecasting, bioinformatics, robotics, and autonomous systems are faced with limitations tied to the conventional computing paradigm.

Research and industry are exploring radical new computing paradigms such as quantum computing and exploring new solutions to yet unresolved problems and challenges—all of which have the potential to bring about a variety of promising new applications. Understanding, mastering, and applying them will empower us to chart the future course of computing. This book explores principles of and potential for some of these paradigms and approaches and examines their current status and future prospects.

Book Preview

The book presents 14 chapters in five parts, each focusing on an emerging area: cloud computing; quantum computing and its applications; computational intelligence and its applications; advances in wireless networks; and blockchain technology and cyber security.

Cloud Computing

Cloud computing fundamentally changed the IT industry and the ways applications are built and deployed. This computing paradigm is now being widely used for a variety of applications by individuals, business, and government. The first chapter, "Cloud Computing: Evolution, Research Issues, and Challenges" presents a brief, yet comprehensive, overview of cloud computing and outlines its key technologies and approaches. The chapter also highlights cloud's challenges and current limitations and discusses several key issues that require further study. Next chapter, "Cloud IoT: An Emerging Computing Paradigm for Smart World," outlines the role of cloud computing and IoT in the emerging smart world and describes a range of applications. It presents a cloud-IoT architecture and briefly describes supporting infrastructure such as edge computing, fog computing, and mist computing. This chapter also discusses challenges and issues in cloud-IoT integration and adoption.

Quantum Computing and Its Applications

Quantum computing is a new radical computing paradigm. It is fast evolving and attracting growing interest as it has potential to solve computationally intensive problems in a range of areas that are unsolvable even by current supercomputers. The chapter "Quantum Computing: Principles and Mathematical Models" explains the quantum phenomena, superposition and entanglement, and explains introductory mathematics that describe working of a quantum computer, such as mathematical notions of single and composite qubits, quantum measurement, and quantum gates and circuits. It also illustrates, with the help of Deutsch Algorithm, supremacy of quantum algorithms over their classical counterparts.

The next chapter, "Quantum Cryptography and Security," beginning with a brief introduction to encryption, provides a brief overview on quantum cryptography. It describes relevant protocols and their applications in cryptography and explains the use of quantum distributive encryption and generation and distribution of quantum keys. It also discusses the effect of noise and spy monitoring over the communication network and error correction through which the destination user can recover the original message from the corrupted data.

Adoption of quantum computing in machine learning presents several new opportunities to solve problems in new smart ways. The Chapter "Quantum Machine Learning Algorithms" presents an overview of recent progress in quantum machine learning and outlines and compares different quantum machine learning algorithms.

Computational Intelligence and Its Applications

In this part, we present three chapters which discuss application of computational intelligence (CI) in three domains. First, in "Computational Intelligence Paradigms in Radiological Image Processing—Recent Trends and Challenges," the authors briefly introduce CI and present an overview on radiological information processing using computational intelligence paradigm and discuss emerging trends. They also describe different stages in radiological image processing and explains the use of CI paradigms based on fuzzy logic, artificial neural networks, and evolutionary computation.

Next, "Computational Intelligence in Agriculture" discusses various uses of CI in agriculture, such estimation and improvement of crop yield, water conservation, soil and plant health monitoring, and plant disease detection. The chapter also describes various remote sensing methods and different wireless communication protocols and machine learning models used in agriculture applications.

Though recurrent neural networks (RNNs) are effective in handling sequential data, they have limitations in capturing the long-term dependencies in the data due to a problem known as vanishing and exploding gradients. Long-and-Short-Term Memory (LSTM), a variant of RNN, overcomes this problem and is very efficient and offers better performance in handling sequential data. The chapter "Long-and-Short-Term Memory (LSTM) Networks: Architectures and Applications in Stock Price Prediction" presents the basic design of LSTM networks and describes their working principles. It discusses and compares six different variants of LSTM models for stock price forecasting. The models were trained and tested on real-world data—the historical NIFTY 50 index of the National Stock Exchange (NSE) of India from December 29, 2014 to July 31, 2020.

Advances in Wireless Networks

Wireless networks are rapidly advancing and finding application in several areas. The 5G revolution aims to merge all communication networks into one ubiquitous global network with a seamless integration of communication services that are transparent to the mobile end users and devices. The chapter "Mobile Networks: 5G and Beyond" reviews main principles of mobile networks and outlines the strategies and challenges in designing 5G mobile networks. In particular, it outlines the concepts, design challenges and key developments in vehicle ad-hoc networks (VANETs) including connected cars, unmanned aerial vehicle (UAV), and low-Earth orbit (LEO) satellites.

Advanced Wireless Sensor Networks (AWSNs) are gaining growing interest among the researchers and practitioners as they have a range of applications like structural health monitoring, precision smart agriculture, smart cities, and smart parking systems. AWSNs, however, present several issues and challenges that need to be addressed. The chapter "Advanced Wireless Sensor Networks: Research Directions" identifies and discusses research areas in AWSNs. It also outlines key principles governing design of mobile networks which are now directly integrated with the 5G cellular networks.

The final chapter in this section, "Synergizing Blockchain, IoT and AI with VANET for Intelligent Transport Solutions," looks at benefits and challenges of embracing the potential of blockchain, IoT and AI with VANET in the context of facilitating intelligent transport solutions. It also examines security aspects of this synergic combination of technologies, showcases novel applications that address UN Sustainable Development Goals (SDGs) and outlines future trends in this area.

Blockchain Technology and Cyber Security

In this part, we feature three chapters focusing on blockchain, cryptocurrencies, and cybersecurity.

Enterprise blockchains have the ability to scale well, are less decentralized than a public blockchain, which matches their use case, and presents fewer potential security issues as they are permissioned networks. The chapter "Enterprise Blockchain," introduces enterprise blockchain and examines a few of its use cases.

The chapter "Blockchain and Cryptocurrencies: Techniques, Applications, and Challenges" describes key elements of blockchain and its features, highlights few applications, and identifies the limitations and challenges.

The final chapter, "Importance of Cyber Security and Its Subdomains," briefly discusses key aspects of cybersecurity and identifies and discusses cybersecurity subdomains. It also identifies emerging threats that need to be addressed further.

The book gives a glimpse of emerging computing technologies and paradigms and identifies some of limitations and challenges that computing presents. It also identifies research issues in the respective areas that require further study. We encourage researchers and developers from multidisciplinary fields to learn from each other and work together to further advance computing and its applications.

Umang Singh
Ashish Seth
San Murugesan

Acknowledgements

Publication of this book wouldn't have been possible without the contribution, support, and cooperation of several people. We acknowledge and appreciate their contributions and support.

First, we would like to thank each one of the chapter authors for enthusiastically contributing to the book, and thereby sharing their expertise, experiences and insights with the readers. We gratefully acknowledge their support and cooperation. We also extend our gratitude to the reviewers who have provided valuable comments on the chapter manuscripts.

Next, the editorial team at Wiley deserves our high commendation for their key roles in publishing this book and in ensuring its quality. In particular, we would like to thank Sandra Grayson, Commissioning Editor; Juliet Booker, Managing Editor; and Becky Cowan, Editorial Assistant for their excellent enthusiasm, support, and cooperation. We would like to thank the staff at Integra for their excellent work on this book which helped to shape and improve the presentation. We highly commend their professionalism and commitment.

Finally, we would like to thank our respective family members for their encouragement, support, and cooperation which enabled us to make this venture a reality and enjoyable.

About the Editors

Umang Singh was Associate Professor at Institute of Technology & Science, Ghaziabad, UP, India. She has been involved with research and academia for more than 17 years. She was renowned for her keen interest in the areas of Mobile Networks, IoT, Edge Computing and Machine Learning. She has published over 80 research papers in reputed journals and conferences like ACM, Elsevier, Inderscience, IEEE, Springer indexed in SCI, ESCI, SCIE, and Scopus.

She served as Guest Editor for special issues of journals including *International Journal of e-Collaborations* (IGI Global, USA, 2020), and *International Journal of Information Technology* (BJIT 2010) and edited six Conference Proceedings, three souvenirs and two books. Dr Umang was on the Editorial Board for several reputed journals including *Inderscience IJFSE* (Switzerland), on the Board of Referees for the *International Journal of Information Technology*, BJIT, Springer and the Technical Programme Committee Member of national and international Conferences. She was a senior member of IEEE and life member of Computer Society (CSI).

San Murugesan is Director of BRITE Professional Services and an Adjunct Professor with Western Sydney University, Sydney, NSW, Australia. He has vast experience in both academia and industry. He is former Editor-in-Chief of the IEEE Computer Society's *IT Professional* magazine and co-editor of several books, including the *Encyclopaedia of Cloud Computing* (Wiley and IEEE), *Harnessing Green IT: Principles and Practices and Web Engineering* (Springer). He guest edited 40 journal special issues and served on editorial boards of several reputed international journals.

He worked as a Senior Research Fellow at the NASA Ames Research Center in California and served as professor of computer science at Southern Cross University in Australia. Prior to these, he worked at the Indian Space Agency in Bangalore in senior roles and led development of onboard microcomputer systems. He is a Distinguished Speaker of ACM and Distinguished Visitor of IEEE Computer Society. He is a fellow of the Australian Computer Society and the Institution of Electronics and Telecommunication Engineers, a Golden Core member of IEEE Computer Society and a Life Senior Member of IEEE. For further details visit www.tinyurl.com/san1bio.

Ashish Seth is an author, consultant, researcher and teacher. He is a Professor at the School of Global Convergence Studies, Inha University, South Korea. He is also a visiting faculty at TSI, Riga, Latvia. He is PhD (Computer Science) in the area of Information Systems from Punjabi University, Patiala, Punjab, India and holds and MPhil(CS) and MCA degree. He has published more than 40 research papers in indexed journals. He has authored four books and several book chapters. He also edited two books and one indexed journal.

He is senior member IEEE, life member CSI and an active member of International societies like IACSIT, IAENG, etc. He is also ACM-Distinguished speaker. He has been involved with research and academia for more than 17 years. He worked at various universities in India and abroad holding different positions and responsibilities. He has organized and participated actively in various conferences, workshops and seminars. He served as Subject Expert under the European Union in the areas of Strategic Specialization. His research interests include Service Oriented Architecture, Optimal Computing, Cloud Computing and Blockchain Technologies. He finds interest in reading and writing articles on emerging technologies (https://www.linkedin.com/in/dr-ashish-seth-877b1116)

About the Contributors

Chapter 1

Neeraj Gupta received the BTech Degree in Computer Engineering and M.Tech in Computer Science from Kurukshetra University and JRN Rajasthan Vidyapeeth (Deemed University) in 1999 and 2006 respectively, and the PhD degree from University School of Information, Communication & Technology at Guru Gobind Singh Indraprastha University, Delh in 2016. From October 1999 to July 2014 he worked as Assistant Professor in Computer Science Engineering in Hindu College of Engineering, Sonepat, India. He is working as Associate Professor with School of Engineering and Technology at K.R. Mangalam University, Gurugram, India since July 2014. He has successfully guided three PhD scholars. He is Professional Member of ACM and life member of Computer Science of India. His research interests are in area of Cloud Computing, Edge Computing, Software Defined Networks, Blockchains, Sensor Networks and Performance Modeling and Analysis of networks.

Asha Sohal received the BTech Degree in Computer Engineering and MTech in Information Technology from Kurukshetra University and Guru Gobind Singh Indraprastha University, Delhi(GGSIPU) in 2003 and 2011 respectively, and pursuing PhD degree from Department of Computer Science and Applications, Kurukshetra University, Kurukshetra. From August 2003 to July 2007 she worked as senior lecturer in Computer Science Engineering in reputed colleges of engineering, India. From August 2008 to July 2011 and July 2011 to July 2013  she worked as Assistant Professor in Computer Science Engineering in KIIT, Gurgaon and Amity University, Gurgaon respectively. She is working as Assistant Professor with the School of Engineering and Technology at K.R. Mangalam University, Gurugram, India since July 2013. Her research interests are in area of Cloud Computing, Fog Computing, Wireless Sensor Networks and MANETs.

Chapter 2

Ruchi Bhatnagar is a dynamic and prominent research scholar and academic having 15+ years of experience in teaching and study at IIMT University, Meerut. Her primary research fields are Networking and Algorithm, ranging from theory to design to implementation. She has published her research in several international and national journals and organized and served international conferences program committees at IIMT University, Meerut.

Paramjeet Rawat is Professor at the IIMT Engineering College, Meerut. She is a reviewer on several reputed international journals at Elsevier, Inderscience, IJAIS, IJARCSEE, TIJCSA and has published her research in more than 25 reputed and referred journals. She had also presented a curriculum on ethical education in Amherst College, USA; where she represented team India and received a letter of appreciation from MLI, USA for her excellent work. She has good analytical and problem solving skills and guided of many PhD scholars.

Dr. Amit Garg is a dynamic and vibrant academic having 19 years of experience. Currently he is working as HoD, Department of CSE, IIMT Engineering College, Meerut. He is a life-time member of CSI, IANEG and ISRD. He is an eminent scholar with his international publications including *SCOPUS* and *ESCI* Journals. Throughout his carrer he has also worked as SPOC for IBM, WIPRO, TCS, and SAP.

Chapter 3

Arish Pitchai is working as Assistant Professor, Department of Computer Science, CHRIST (Deemed to be university), Bengaluru, India. He is also associated as Quantum Scientist, Quantum Machine Learning Lab, BosonQ Psi Pvt. Ltd, Bhilai, India. After completing his PhD in quantum game theory, Arish Pitchai worked for a while as an associate consultant in Atos, Quantum R&D. Then he joined CHRIST (Deemed to be University) as an Assistant Professor.

Thiruselvan Subramanian works as Assistant Professor, Department of Computer Science, Presidency University, Bengaluru, India. He pursued his PhD in cloud computing technology and his research interests include computer vision and quantum computing.

Chapter 4

Vandana Niranjan is working as Professor in the Department of Electronics and Communication Engineering at Indira Gandhi Delhi Technical University Delhi, India. She graduated in the year 2000 and received her BE degree in Electronics and Communication Engineering from Government Engineering College (now University Institute of Technology of Rajiv Gandhi Proudyogiki Vishwavidyalaya) Bhopal. In the year 2002, she received her MTech degree from the Department of Electronics and Communication Engineering at Indian Institute of Technology (I.I.T) Roorkee with VLSI Design as specialization. In the year 2015, she was awarded her PhD degree in the area of Low Voltage VLSI Design from University School of Engineering & Technology, GGSIP University, Delhi. She has 20 years' teaching and research experience at Indira Gandhi Delhi Technical University, Delhi. Her areas of interest includes MOSFET body bias techniques and low-voltage low-power analog CMOS circuit design. She has several publications to her credit in various international journals and conferences and book chapters.

Anukriti has recently completed her Masters in technology in the area of VLSI and Chip design at Indira Gandhi Delhi Technical University, Delhi, India. She graduated in the year 2017 and received her BTech degree in Electronics and Communication Engineering from Guru Gobind Singh Indraprastha University, Delhi. She is currently working as a Research Intern under the Openlabs Programme at CERN, Geneva where she is pursing her research career in Quantum Machine Learning and Computing. She is an aspiring researcher and scientist and has published in indexed international conferences when her main area of focus was Machine Learning, Deep learning, Dark Matter Physics and Applied Cosmology.

Chapter 5

Renata Wong received her MA degree in Sinology, as well as BSc. and MSc. degrees in Computer Science from Leipzig University, Germany, in 2008, 2011, and 2013, respectively. She holds a PhD in Computer Science (Quantum Computing) from Nanjing University, PRC, and is currently a postdoctoral research fellow at the Physics Division, National Center for Theoretical Sciences in Taipei, Taiwan, R.O.C, working in quantum computation and information. Her main research fields are quantum computation and information (especially quantum algorithms), protein structure prediction, foundations of physics, and linguistics. Up to date, she has published over two dozen journal papers, conference papers and book chapters in English, Chinese and German. Within the field of quantum computing, she has, among others, developed and successfully simulated two quantum algorithms for protein structure prediction. In the area of physics, her contributions are in the research on the logical consistency of physics theories, with focus on the special and the general theory of relativity and quantum mechanics. In addition to English and her native Polish, she is fluent in Mandarin Chinese, German, and Russian. The current affiliation is: Physics Division, National Center for Theoretical Sciences, Taipei, Taiwan.

Tanya Garg is a student at the Indian Institute of Technology, Roorkee, India, pursuing a bachelor's degree in Engineering Physics with a minor in Computer Science and Engineering. Interested in the fundamental of physics and computer science, she intends to pursue the field of quantum computing and the related areas of quantum communication and quantum algorithms due to their interdisciplinary nature and the technological potential they propose. She has conducted undergraduate research at renowned laboratories and is the recipient of some prestigious scholarships, namely the DAAD WISE scholarship and the Charpak Lab Scholarship.

Ritu Thombre completed her BTech from Visvesvaraya National Institute of Technology in Computer Science and Engineering. She recently joined CITI India as Technology Analyst. Her primary interests are Cryptography, Machine Learning, Quantum Mechanics and Astrophysics. She likes to design new machine learning algorithms to solve various problems that have real world applications. She is deeply fascinated by applications of Quantum Computing in Cryptography and Deep Learning. Currently we live in a "vacuum-tube" era of Quantum Computers and hopes to contribute to the advancement of Quantum Computing to achieve "quantum-supercomputer" era someday.

Alberto Maldonado Romo is a PhD student in Computer Science at the Center for Computing Research, Instituto Politecnico Nacional, Mexico, where he has collaborated with the Fermilab and CERN research centres on the GeantV project, and did his Master's thesis on quantum image processing. He a technical reviewer of the book *Quantum Computing with Silq Programming*. Alberto is an administrator for the non-for-profit company Quantum Universal Education, he has conducted workshops and webinars on introduction to quantum computing, designed comics introducing beginning quantum computing concepts, and created tutorials in five different quantum programming languages.

Niranjan PN, a current Master's student of Physics at the University of Cologne, is also a quantum developer in a promising startup - BosonQ Psi Pvt.Ltd in India. He has more than two years of experience in the field of quantum computing. His area of interest lies in fault-tolerant quantum hardware and variational quantum algorithms. He is currently working on different projects one of which is covid detection using quantum machine learning. He has presented two lectures; one in Kongunadu Arts and Science College, Coimbatore, India on quantum computing as a beginner and one at the Government College of Engineering, Thanjavur under the Faculty Development Program on Quantum Computing organized by AICTE Training and Learning (ATAL) Academy. He is also a freelance consultant on how to start one's career in quantum computing.

Pinaki Sen is a final year Electrical Engineering undergrad at National Institute of Technology, Agartala, India. His interest lies primarily in the domain of Quantum Computing and Machine Learning. He has research experience in Quantum Machine Learning, Quantum Error Correction, Quantum-dot Cellular Automata and related fields. He has previously worked as a research intern at the Indian Statistical Institute, Kolkata, India.

Mandeep Kaur Saggi is presently pursuing a PhD in Computer Science Engineering at Thapar Institute of Engineering & Technology, Patiala. She is working on multi-level ensemble modelling for predicting the Reference Evapotranspiration (ETo) and Crop Evapotranspiration for Crop Water Requirement, Irrigation Water Requirement, and Irrigation Scheduling using Artificial Intelligence, Deep Learning, and Machine Learning techniques. She completed her MTech in Computer Science Engineering at D.A.V University, Jalandhar. She completed her BTech in Computer Science & Engineering from Punjab Technical University. Her area of interest lies in Deep Learning, Machine Learning, Big Data Analytics, Quantum Computing, & Cloud Computing.

Amandeep Singh Bhatia is working as a Postdoc in Institute of Theoretical Physics at University of Tubingen, Germany. He is developing quantum computational models and quantum algorithms with new artificial intelligence methods for physics, and novel quantum machine learning techniques. He completed his PhD degree in the realm of Quantum Computation and Information at the Computer Science & Engineering Department at Thapar University in July 2020. He has more than six years of work experience in quantum computing and technologies. He contributed to the progress of quantum automata theory and developed quantum computational models for Biology, Chemistry and Tensor network theory. Presently, he is contributing to the progress of quantum machine learning. He received his BTech and MTech degrees in Computer Science & Engineering in 2010 and 2013, respectively. Up to date, he has published over two dozen journal papers, conference papers and book chapters on several aspects of quantum computation and information.

Chapter 6

Anil B. Gavade is a Associate Professor at the KLS Gogte Institute of Technology, Belagavi, Karnataka, India, in the Department of Electronics and Communication Engineering. He received a BE degree in Instrumentation Engineering from Karnataka University, Dharwad, an MTech in Digital Electronics and PhD in Electrical and Electronics Engineering from Visvesvaraya Technological University, Belagavi. His main research interests include Computer Vision and Machine Learning Applications to Biomedical Imagery and Satellite Imagery.

Rajendra B. Nerli is a Professor and Head, Department of Urology, JN Medical College, KLE Academy of Higher Education and Research (Deemed-to-be-University), Belagavi. Dr Nerli has published over 300 research articles in peer-reviewed indexed journals. He has carried out clinical experimental and field research in the areas of urology, artificial intelligence and medical sciences. As a research mentor, he has guided a number of students leading to MCh and PhD Degrees. He is the Director of Clinical service at KLES Dr Prabhakar Kore Hospital & MRC, Belagavi.

Dr Ashwin S. Patil is presently Professor and Head, Department of Radio-diagnosis, JNMC Belagavi. He completed his Medical School (MBBS) in 1994 and MD-Radiology in 1997 from Jawaharlal Nehru Medical College, Belagavi and Karnataka University, Dharwad. He has over 27 years of teaching experience and has mentored and guided over 50 postgraduates. He has authored over 17 publications, four of which are published in leading international journals and has presented over 66 presentations (paper, e-poster). He also holds and has held several posts, prominent among them being Board of Studies Member for KAHER since 2014 and BLDE University (2016–2019), Affiliation Inquiry Committee member for

Goa University (2016–2017) and Life members of IRIA (Indian Radiology and Imaging Association) and IMA (Indian Medical Association). His subspecialty interests include CT and MRI reporting. He is currently involved in research activities on Artficial Intelligence and tuberculosis in collaboration with Thomas Jefferson University, USA.

Shridhar Ghagane is currently working as Research Scientist (R&D) at Urinary Biomarkers Research Centre, KLES Dr. Prabhakar Kore Hospital & Medical Research Centre, Belagavi, India. His area of expertise is in Medical Biotechnology, Urologic-oncology, Cancer Biomarkers and Artificial Intelligence. He has published over 150 research articles in national and international peer-reviewed journals. Currently, he is supervising two PhD candidates at the KLE Academy of Higher Education and Research, Belagavi.

Venkata Siva Prasad Bhagavatula is working as Principal Systems Engineer with Medtronic Innovation and Engineering center, Hyderabad, India. He completed his BE degree in Instrumentation Engineering from Karnataka University, Dharwad and completed his Master of Science (Online) in Data science from Liverpool John Moores University, UK. He holds a patent, launched three new products in various roles in Research and Development in the Healthcare industry. He has 18 years of experience in the areas of hardware and systems engineering. His main area of work is in systems engineering. His main interests include medical devices, data science and machine learning applications to medical imaging.

Chapter 7

Hari Prabhat Gupta is an Assistant Professor in the Department of Computer Science and Engineering, Indian Institute of Technology (BHU) Varanasi, India. Previously, he was a Technical Lead at Samsung R&D Bangalore, India. He received his PhD and MTech degrees in Computer Science and Engineering from the Indian Institute of Technology Guwahati in 2014 and 2010 respectively; and his BE degree in Computer Science and Engineering from Govt. Engineering College Ajmer, India. His research interests include the Internet of Things (IoT), Wireless Sensor Networks (WSN), and Human-Computer Interaction (HCI). Dr. Gupta has received various awards such as Samsung Spot Award for outstanding contribution in research, IBM GMC project competition, and TCS Research Fellowship. He has guided three PhD thesis and five MTech dissertations. He has completed two sponsored projects and has published three patients and more than 100 IEEE journal and conference papers.

Swati Chopade received her MTech Degree in Computer Science and Engineering from VJTI, Mumbai, India. Presently, she is pursuing a PhD at the Department of Computer Science and Engineering, IIT (BHU) Varanasi. Her research interests include machine learning, sensor networks, and cloud computing.

Tanima Dutta is an Assistant Professor in the Department of Computer Science and Engineering, Indian Institute of Technology (Banaras Hindu University), Varanasi, India. Previously, she was a researcher in TCS Research & Innovation, Bangalore, India. She received a PhD from the Dept. of Computer Science and Engineering, Indian Institute of Technology (IIT) Guwahati in 2014. Her PhD was supported by TCS (Tata Consultancy Services) Research Fellowship and she received SAIL (Steel Authority of India Limited) Undergraduate Scholarship for perusing her BTech degree. Her research interests include (major) Deep Neural Networks, Machine Learning, Computer Vision, and Image Forensics and (minor) Human-Computer Interaction (HCI) and Intelligent Internet of Things (IIoT).

Chapter 8

Jaydip Sen has around 28 years of experience in the field of networking, communication and security and machine learning, and artificial intelligence. Currently, he is associated with Praxis Business School as Professor and the Head of the School of Computing and Analytics. He is also a visiting Professor to XLRI Jamshedpur and also an IBM ICE (Innovation Center for Education) Subject Matter Expert (SME). His research areas include security in wired and wireless networks, intrusion detection systems, secure routing protocols in wireless ad hoc and sensor networks, privacy issues in ubiquitous and pervasive communication, machine  learning, deep learning and artificial intelligence. He has more than 200 publications in reputed international journals and refereed conference proceedings, and 22 book chapters in books published by internationally renowned publishing houses, like Springer, CRC press, IGI-Global, etc., and four books published by reputed internal publishing house. He is a member of ACM and IEEE. He has been listed among the top 2% most cited scientists in the world as per a study conducted by Stanford University which has been published in a paper in PLOS ONE journal in September 2020.

Sidra Mehtab has completed her BS with honors in Physics from Calcutta University, India in 2018. She has an MS in Data Science and Analytics from Maulana Abul Kalam Azad University of Technology (MAKAUT), Kolkata, India in 2020. Her research areas include Econometrics, Time Series Analysis, Machine Learning, Deep Learning, and Artificial Intelligence. Ms. Mehtab has published ten papers in reputed international conferences and two papers in prestigious international journal. She has won the best paper awards in two international conferences – BAICONF 2019, and ICADCML 2021, organized at the Indian Institute of Management, Bangalore, India in December 2019, and SOA University, Bhubaneswar, India in January 2021. Besides, Ms. Mehtab has also published two book chapters in two books published by IntechOpen, London, UK. Seven of her book chapters have been published in the December 2021 volume by Cambridge Scholars' Press, UK. Currently, she is working as a Data Scientist with an MNC in Bangalore, India.

Chapter 9

Pavel Loskot joined the ZJU-UIUC Institute in Haining, China, in January 2021 as the Associate Professor after spending 14 years at Swansea University in the UK. He received his PhD degree in Wireless Communications from the University of Alberta in Canada, and the MSc and BSc degrees in Radioelectronics and Biomedical Electronics, respectively, from the Czech Technical University of Prague in the Czech Republic. He is the Senior Member of the IEEE, Fellow of the Higher Education Academy in the UK, and the Recognized Research Supervisor of the UK Council for Graduate Education. He has been involved in numerous telecommunication engineering projects since 1996. His current research interests focus on the problems involving statistical signal processing and importing methods from Telecommunication Engineering and Computer Science to other disciplines in order to improve the efficiency and information power of system modeling and analysis.

Chapter 10

Richa Sharma completed her MTech (Computer Science & Engineering), MS, and BSc from Guru Nanak Dev University (GNDU), Amritsar. Ms. Sharma has 4+ years of teaching experience. She has expertise in CS subjects like Computer Networks, Data Structures and Algorithms and has taught these subjects to engineering students. Currently, she is pursuing her doctoral degree in the area of Wireless Sensor Networks. Also, she is a keen researcher and up to 2020 has published 17 research papers. She has presented her research work at international  level in IEEE conferences and has had her work published in SCI, SCIE and Scopus Indexed International Journals. She has also contributed voluntarily as a reviewer for reputed International Journals and IEEE Conferences. Her primary areas of interests include the Wireless Sensor Networks and Evolutionary Techniques.

Chapter 11

Sheetal Zalte is assistant professor in Computer Science Department at Shivaji University, Kolhapur, India. She pursued her a BSc from Pune University, India and MSc from Pune, India. She earned her PhD in Mobile Adhoc Network at Shivaji University. She has 14 years of teaching experience in computer science. She has published 20+ research papers in reputed international journals and conferences including IEEE (also available online). She has also authored book chapters with Springer, CRC. Her research areas are MANET, VANET, Blockchain Security.

Vijay Ram Ghorpade has completed PhD in Computer Science and Engineering from Shri Guru Govindsinghji Institute of Engineering and Technology (An Autonomous Institute of Govt. of Maharashtra), Nanded in 2008. He has more than 30 years of teaching experience at different levels. Presently he is working as the Principal, Bharati Vidyapeeth College of Engineering, Kolhapur. His area of interest is in Internet Security, Mobile Ad hoc Networks, Authentication, Cloud Security, Block Chain, etc. So far he has published more than 76 papers in peer reviewed national and international

journals of repute. He has presented more than 69 papers in national and international conferences organized by IEEE, ACM, CSI. He is a member of various professional bodies such as, ACM, IEEE Computer Society, ISTE, CSI, CRSI, ISACA. He received '*Best Engineering Principal Award*' from ISTE-Maharashtra-Goa Section in 2015. He has filed five patents, and two of which have been published in the public domain.

Rajanish K. Kamat is currently holding the position of Dean, Faculty of Science & Technology in addition to Professor in Electronics and Head of the Department of Computer Science at Shivaji University, Kolhapur a NAAC A++ accredited HEI. He is also Member of Management Council, Academic Council and Senate of the University. Until recently he also served as Director, IQAC (2014–2020) and Director, Innovation, Incubation & Linkages (2018–2020). He has to his credit 200+ publications in journals from reputed publishing houses such as IEEE, Elsevier, Springer in addition to 16 reference books from reputed international publishers such as Springer, UK and River Publishers, Netherlands and exemplary articles on ICT for Encyclopedia published by IGI. He is a Young Scientist awardee of Department of Science and Technology, Government of India under Fast Track Scheme. Dr. Kamat is also currently working as an Adjunct Professor in Computer Science for the reputed Victorian Institute of Technology, Melbourne, Australia.

Chapter 12

Ashish Seth is an author, consultant, researcher and teacher. He is a Professor at School of Global Convergence Studies, Inha University, South Korea and is deputed at Inha University Tashkent, Uzbekistan. He is PhD (Computer Science) in the area of Information Systems from Punjabi University, Patiala, Punjab, India and holds MPhil(CS) and MCA degree. He worked at various universities in India and abroad holding different positions and responsibilities. He is senior member IEEE, life member CSI and an active member of International societies like ACM, IACSIT, IAENG, etc. He has delivered many invited talks and also serving as ACM-Distinguished speaker under ACM-DSP program. He has been a consultant with many projects and associated with project granted by Indian government and European Union. His research interests include Service Oriented Architecture, Cloud Computing and

Blockchain Technologies. He finds interest in reading and writing articles on emerging technologies.

Kirti Seth is researcher and academic. She has a PhD (Computer Science and Engineering) in the area of Component Based Systems from Department of Computer Sciences and Engineering, from AKTU, Lucknow, India and an MTech (Computer Science) from Banasthali Vidyapeeth, Banasthali, Rajasthan, india. She also holds an MSc (CS) degree and has been into research and academia for the last 16 years. She is presently working as Associate Professor at Inha University in Tashkent, Tashkent, Uzbekistan. She has published more than 40 research papers in reputed journals like ACM, Springer and Elsevier and authored four books. She has been participating

and organizing seminar, conferences, workshops, expert lecturers and technical events to share knowledge among academics, researchers and to promote opportunities for new researchers. She has provided training programs for students and faculties on various areas of computer science including Google's techmaker event-2018. She has given keynote talks at many international conferences. Her current research interests include Service Oriented Architecture, Bio Inspired Optimizations, Neural Networks and Component Based Systems. She has been awarded "Young Scientist in Software Engineering-2017" in ARM 2017. She also received "Most Promising Women Educationist" of the year award in India Excellence Summit 2017, on 19th August 2017.

Himanshu Gupta is working as an Associate Professor in the reputed international university Amity University Uttar Pradesh, Noida in India. He completed all his academic as well as professional education from the reputed Aligarh Muslim University, Aligarh (Uttar Pradesh), India. He has visited Malaysia, Singapore, Thailand, Cambodia, Vietnam, Indonesia, Hong Kong, Macau, China and United Arab Emirates for his academic and research work. He has delivered many Technical Sessions on "Network Security & Cryptography" in the field of Information Technology in various reputed International Conferences, World Summit and other

foreign universities as an Invited Speaker. He has more than 75 research papers and articles in the field of Information Technology, which have been published in various reputed Scopus and other indexed conference proceedings and journals. He has written a number of books in the area of Information Security, Network Security and Cryptography as a main author published by reputed national and international publishers.

He has successfully filed a number of patents in the area of Network Security and Cryptography as an Inventor, which have been published in the "International Journal of Patents" by Patent Department, Govt. of India. He has delivered Online IT Lectures as an Invited Speaker to students of 16 African Countries under the e-Pan African Project sponsored by Govt. of India at Amity University, Noida.

Chapter 13

Snehlata Barde is working as Professor at MAT'S University, Raipur, (C.G.). She received her PhD in Information Technology and Computer Applications in 2015 from Dr. C. V. Raman University Bilaspur, (C.G.). She obtained her MCA from Pt. Ravi Shankar Shukla University, Raipur, (C.G.) and MSc (Mathematics) from Devi Ahilya University Indore, (M.P.). Her research interest includes Digital Image Processing and its Applications in Biometric Security, Forensic Science, Pattern Recognition, Segmentation, Simulation and Modulation, Multimodal Biometric, Soft Computing Techniques, cyber crime, IoT. She has published 61 research papers in various international and national journals and conferences. She has attended 36 seminars, workshop and training programs, she has published six book chapters. She has 22-years' teaching experience from GEC Raipur, NIT Raipur, SSGI Bhilai. She has reviewed the translated files of the course Cloud Computing offered by IIT Kharagpur in Marathi language.

Chapter 14

Parag H. Rughani obtained his Master's Degree in Computer Applications and PhD in computer science from Saurashtra University. He is currently working as an Associate Professor in Digital Forensics at the National Forensic Sciences University, India. He has more than 16 years of experience in academia and has published more than 16 research articles in reputed international journals. He has delivered more than 34 Expert Talks at various levels. His research interests are broadly in the area of Cyber Security and Digital Forensics. He is currently working on Machine Learning, Memory Forensics, Malware Analysis and IoT Security and Forensics.

Part 1

Cloud Computing

1

Cloud Computing

Evolution, Research Issues, and Challenges

Neeraj Gupta and Asha Sohal

School of Engineering and Technology, K.R. Mangalam University, Gurugram

1.1   Introduction

A computing process requires resources like processors, memory, network, and software. The traditional computing model for IT services requires investing in the computing infrastructure. "On-premise" solution requires that you purchase and deploy required hardware and software at your premise. Such a solution involves a capital expenditure on the equipment and recurring operational spending on the maintenance and technological refreshes required from time-to-time. Another possible solution can be co-location facilities where the facility owner can provide services like power, cooling, and physical security. The customer needs to deploy its server, storage, and other equipment necessary for the operation. This solution reduces the capital expenditure and increases the operational cost as per the service-agreement agreed upon for hiring the services. Cloud computing aggregates various computing resources, both hardware and software, such that they are viewed as one large pool and accessed as utility services. The word *utility* refers to hire up the resources until the demand exists and service provider charges for resource usage. The term "pay-per-use" or "pay- as-you-go" is used to represent cloud computing's commercial aspect.

Most users define cloud computing as IT services located somewhere on the cloud, where the cloud presents data centers' location. It is essential to mention here that co-location data centers are off-premises and private clouds are typically on-premise. The National Institute of Standard and Technology (NIST) [1] define cloud computing as "Cloud Computing is a model for enabling ubiquitious, convenient, on-demand network access to a shared pool of configurable computing resources (e.g., networks, servers, storage, applications, and services) that can be rapidly provisioned and released with minimal management effort or service provider interaction. This cloud model is composed of five essential characteristics, three service models, and four deployment models." The definition essentially means that the cloud computing processes are automated, dynamic, and transparent to ensure minimal human intervention.

This chapter is organized in the following way: Section 1.2 discusses various technologies that contributed to the evolution of cloud computing. Section 1.3

Emerging Computing Paradigms: Principles, Advances and Applications, First Edition.
Edited by Umang Singh, San Murugesan and Ashish Seth.
© 2022 John Wiley & Sons Ltd. Published 2022 by John Wiley & Sons Ltd.

explains the characteristics of the cloud, service models, and cloud deployment models. Section 1.4 illustrates various research issues and challenges that confront cloud computing. Section 1.5 describes emerging trends and research challenges associated with them.

1.2 Evolution of Cloud Computing

Cloud computing involves hardware, software processes, and networking. Figure 1.1 illustrates multiple technologies that contribute to the initiation, development, and management of cloud computing services. This section briefly describes key technologies such as grid computing, utility computing, ubiquitous computing, service-oriented architecture, and virtualization that form the foundation pillars for cloud computing.

1.2.1 Grid Computing

It is the form of loosely coupled distributed computing. The resources associated with heterogeneous systems are remotely connected. The system can share each other resources transparently. The system users can access and utilize the resources, including processing power, memory, and data storage. The grid is more involved with huge, non-interactive tasks where the computation can occur independently without communicating intermediate results between the processor [2]. Grid computing helps the service providers to create a huge pool of resources without investing in large and expensive mainframes.

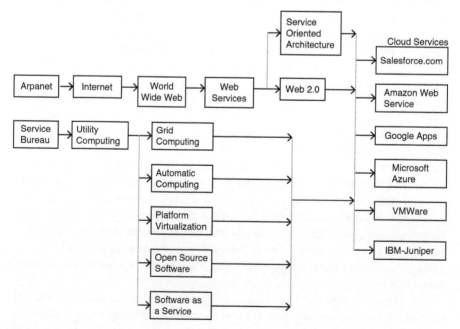

Figure 1.1 Evolution of Cloud Computing (Arockiam, Monikandan and Parthasarathy, 2017).

1.2.2 Utility Computing

Utility computing is a service provisioning model in which a service provider makes computing resources and infrastructure management available to the customer as needed and charges them based on usage rather than a flat rate. It is a business model where customers pay utility company/service providers to access and utilize the computing as per their requirements. The computing resources can be dynamically acquired and subsequently released by the customer. The pay-as-you-go model helps in better resource management of pooled resources offered by the service provider. The customer can save the capital expenditure and recurring operational expenditure toward infrastructure to run the business [4].

1.2.3 Ubiquitous Computing

Mark Weiser first proposed the idea of ubiquitous computing in 1988 [5]. The authors' impression was to integrate computing into the everyday life of people transparently and seamlessly. It was envisioned that computing devices would come into various sizes, each developed to perform a particular task. The system will consist of specialized hardware and software that can communicate using wired media, radio waves, and infrared waves. The work laid the foundation of context-aware systems and innovative applications [6]. Pervasive computing is the convergence of mobile computing, ubiquitous computing, consumer electronics, and the Internet.

1.2.4 Service-Oriented Architecture

Service-Oriented Architecture (SOA) is a software design process that enables software components to access other software components' functionality through a communication protocol over a network. The software components, viewed as services, offer a discrete functionality that can be accessed, updated, and acted upon remotely by other services to fulfill a task. The design promotes the loosely coupled distributed software components that can communicate with each other using a message-passing communication model. The SOA philosophy is independent of vendor, product, and technology [7]. Web services are based on the concept of SOA. W3C defines web services as a "software system designed to support interoperable machine-to-machine interaction over a network. It has an interface described in a machine-processable format specifically web service description language (WSDL). Other systems interact with the web service in a manner prescribed by its description using Simple Object Access Protocol (SOAP) messages, typically conveyed using HTTP with an XML serialization in conjunction with other web-related standards" [8].

Figure 1.2 illustrates the Find-Bind-Execute paradigm that defines the working of the web services. The service provider publishes the software components as service in a common registry. The service consumer can find the particular service using WSDL. Once the service is located service requester can invoke or initiate the web service at runtime.

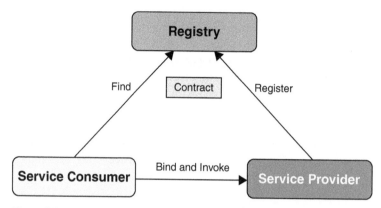

Figure 1.2 SOA's find-bind-execute paradigm (Qusay H. Mahmoud, 2005).

1.2.5 Virtualization

Personal computing endorsed the single operating image per machine. The operating system and hardware are tightly coupled with each other. This kind of arrangement does not fully utilize the power of underlying hardware. Whenever multiple applications are initiated, resource conflicts can occur, leading to under performance of the individual applications. Virtualization is a framework that divides physical computing components into numerous virtual resources. It relies on the software to manage various resources virtually and distribute to multiple applications flexibly and adhering to specific requirements. The goals of the virtualization architecture are equivalence, resource control, and efficiency [10]. The large pool of computing resources, including server, memory, network, operating systems, and applications are virtualized in data centers. These virtualized resources are offered to the end-users as metered services.

1.3 Cloud Computing Essentials

The US National Institute of Standard and Technology has defined cloud computing through five essential characteristics, three service models, and four deployment model [1].

1.3.1 Characteristics On-Demand Self-Service

In comparison to on-premise solutions and co-location facilities, the resources are provisioned automatically. There is no human interaction whenever the customer wants to access the resources offered by the service provider. The cloud computing processes involved in provisioning are autonomous, which helps cut the cost and adapt to dynamic changes transparently while hiding the internal complexity to maintain such a system.

1.3.1.1 Broad Network Access

Services are delivered by the service provider and availed by the customer through a standard communication protocol. The applications can be accessed irrespective of the type of device. The only requirement is that machines should be online.

1.3.1.2 Resource Pooling

Various computing resources, including server time, storage, network, operating system, middleware, and application are aggregated and offered as services through the virtualized workspace. Multiple customers, also referred to as tenants, can share these resources in an isolated manner. There is a sense of location independence where the user has no control over the way data is stored and retrieved from the data centers.

1.3.1.3 Rapid Elasticity

The resources can be scaled up or released automatically anytime as per the requirement raised by the customer. The dynamic provisioning of the resources helps the service providers optimize the utilization of the data center resources.

1.3.1.4 Measured Services

The resources availed by the customer are usually metered services. The metered service supports the "pay-per-use" or "pay-as-you-go" models. It helps both the service providers and customers to monitor resource usage.

1.3.2 Service Models

The resources in cloud computing are managed in a centralized manner. The virtualized resources offered as service to the customer include infrastructure, operating system, middleware, runtime system, and software applications. The cloud service model is represented as a stack of these services and divide broadly into three components. The infrastructure components include the computing model, storage model, and network model and forms the stack's first component. The operating system, middleware, and runtime systems are usually taken as the second layer of the stack. Various software applications are running on the top of the stack form the third layer.

1.3.2.1 Infrastructure as a Service

The on-demand computing services, which involve computation, storage, and network comes under Infrastructure as a Service (IaaS). The services sometimes extended up to the operating system level. Instead of buying infrastructure, the organizations with low capital purchase these services as fully outsourced services on demand. Organizations can deploy their operating systems, middleware, runtime environment, databases, and applications. Amazon EC2, Amazon S3 bucket, Microsoft Azure, and Rackspace are some of the examples offered as IaaS.

1.3.2.2 Platform as a Service (PaaS)

It involves the customized environment offered by the service provider in terms of infrastructure and customized software environment. These services facilitate application development without undergoing the complexities of managing the infrastructure and runtime system. Aneka, MapReduce, and Google AppEngine are a few of the PaaS service providers.

1.3.2.3 Software as a Service (SaaS)

These are the readily developed applications to meet the specific requirements of the users. The applications can be accessed using web interface. The entire stack is managed by the service providers. Gmail, Outlook 365, Saleforce.com, and Box.net are few of the SaaS providers.

1.3.3 Deployment Models

The deployment models define the procurement strategies for availing the cloud services. The four deployment models can be characterized by the service providers' owner, scale, and service type (see Figure 1.3).

1.3.3.1 Public Cloud

The services are provisioned for open use by the general public. Such clouds are mostly managed and operated by a business organization, academic organizations,

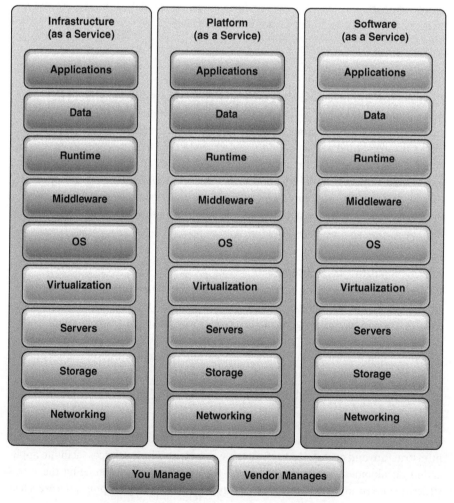

Figure 1.3 Cloud computing service model (Terkaly and Villalobos, 2013).

and the government. The data centers are off-premise entities. Amazon AWS, Rackspace Cloud Suite, and Microsoft Azure are some of the public cloud service providers.

1.3.3.2 Private Cloud
In the private cloud, the services are being provisioned and consumed by a particular organization's members. The data center is owned, managed, and operated by the same organization or hired from some third party. The data center can be either on-premise or off-premise. Eucalyptus, Ubuntu Enterprise Cloud, and VMware Cloud Infrastructure are some of the players in the private cloud.

1.3.3.3 Community Cloud
The services are provisioned for the organizations that have collaborated on a joint project. The cloud can be located either on-premise or off-premise. Open Cirrus and Microsoft government community clouds are a few examples of the community cloud. The data center could be owned and managed usually by both organizations.

1.3.3.4 Hybrid Cloud
In the hybrid cloud, the services can be provisioned and offered from distinct cloud infrastructure. The cloud may change over time based on the usage of resources. Organizations can extend their resources by utilizing both on-premise resources and cloud resources. Cloud bursting is a temporary arrangement to deal with the sudden surge in resources demanded by an application running in a private cloud. The additional resources are hired from the public cloud for the period of burst time.

1.4 Research Issues and Challenges in Cloud Computing

In the last decade, cloud computing has evolved as a mature technology. However, several issues and challenges remains to be satisfactorily addressed.

1.4.1 Resource Management

Resource management is the process of effective delegation of resources and services among multiple users. The cloud resource can be dynamically provisioned and released by the clients making resource management in the cloud a complex task. From the cloud computing perspective, process management is divided into two phases: Ab-initio resource assignment and Period resource management [12]. Some of the prominent open research areas that need to be addressed are discussed below.

1.4.1.1 Risk Analysis
The service level agreement defines the fulfillment of Quality of Service parameters based on specific metrics for the customer's services. Factors like inadequate resources, failure of virtual machines, network failure, and non-responsiveness of applications lead to violation of service level agreements. Since the resources are dynamically provisioned, there is the requirement to define risk management processes that will enable service providers to optimize their resources.

1.4.1.2 Autonomic Resource Management

Elasticity is the essential feature of cloud computing. The resources can be scaled up and scaled down with time. There is a need to address issues like load management, virtual machine migration, and network latency. The design of an intelligent resource management process that can self-manage the elasticity in data centers is an area that is actively investigated.

1.4.1.3 Standardization

Various cloud service providers have different standards to evaluate the service level agreements. Heterogeneity and incompatibility between services of the various cloud service providers is an area that needs thorough investigation. The idea of standardization and benchmarking should be looked upon as a step toward a customer-oriented market.

1.4.1.4 Hybrid and Multi-Cloud Scenarios

Hybrid clouds are the way forward to address the resource crunch faced by academic institutions, small and medium IT companies, and government organizations. The available literature defines a mechanism for leasing the additional resources from the public cloud for the cloud burst time. The hybrid clouds and multi-clouds are seen as supporting elements and not as long-term solutions. Scheduling algorithms, fluctuations during the workflow, and reliability issues can be active research areas. Authors in (Fard et al., 2020) have provided a systematic review of the literature concerning resolution allocation mechanism contributed in the last decade.

1.4.2 Security Challenges

Like traditional computer networks, cloud computing is subjected to security concerns involving confidentiality, integrity, availability, and privacy. Increased usage of Internet and mobile applications has contributed toward an increase in cyber offenses. Various security issues reported in the literature [16, 17] are closely related to each other. Below we discuss different security challenges in cloud computing that needs to be addressed.

1.4.2.1 Trust Issues

Public clouds are extensively customer-driven and user-oriented. The multi-tenant architecture increases the probability of attack from inside the data center. Cloud services are accessible through the Internet. Online attackers can compromise these networks. The service provider must address the issue of identification, authorization, authentication, and access management. The service level documents should include the issues of trust management and future activities.

1.4.2.2 Storage

Location independence is one of the characteristics of cloud computing. The user lacks control over his/her data and is unaware of the data center's location. The service provider deploys data warehouse to manage the data of its customers. Metadata is data about data. It is often maintained for operational reasons. Metadata can be exposed to attackers during the transfer of data over the network, issue of confidentiality, integrity, and availability are the main challenges that need to be addressed.

1.4.2.3 Application Security

Software applications are offered as a service through SaaS model. Platforms like play store are the marketplace for such products. Once deployed on the system, such applications can steal the data, spread malware, tamper the device, lead to repudiation attacks, and initiate denial-of-service. It is crucial to monitor and weed out malicious programs from the cloud environment.

1.4.2.4 Legal Framework

The data center location could be located in different legal jurisdictions, maybe other states or even countries. In case of a security breach, the customer needs to comply with the legal framework prevalent at a data center location. Users need to take care of the site, country, and city when subscribing to the cloud service. The investigative framework to control cloud-related cyber crimes is still in the nascent stage. It is a multidisciplinary area of research that needs active attention.

1.4.3 Green Cloud Computing

Cloud computing has offered several advantages in providing software application services, middleware services, and platform services. The services are managed through the large data centers that constitute computing infrastructure (servers, storage, and networks) and physical electrical infrastructure. According to one estimate, the world's data centers consumed 205 terawatt-hours of electricity in 2018. The massive consumption of energy and heat generated by data centers adds to carbon footprints. Green cloud computing aims to reduce carbon emission through proper utilization of computing resources, disposal of hazardous computing devices, and reducing electricity consumption [18]. To minimize the energy consumption following areas can be pursued.

1.4.3.1 Schedule Virtual Machines

Virtual machine scheduling refers to running the virtual machines on different physical servers in the data center to optimize the computing infrastructure. Power-Aware virtual machine algorithms aim to consolidate virtual machines to low utilized physical devices and dynamically shut down other servers to save energy. Various solutions based on a greedy algorithm [19], genetic algorithm [20], particle swarm optimization [21], and SDN-Cloud [22] has been proposed in recent years.

1.4.3.2 Minimizing Operating Inefficiencies

There are plenty of dependencies that are required to run a simple application in the cloud environment. These dependencies make the system resources hungry and simultaneously more energy consumption to run heavy applications. The containers are lightweight stand-alone packages that include the application and its associated dependencies. The concept of containerization offers the advantages of fast deployment, reconfiguration, interoperability, and portability. The lightweight virtualization technologies such as Docker, Kubernetes, and CoreOS have emerged as popular alternatives. These lightweight containers are reported to perform faster than the traditional virtualized technology [23].

1.4.3.3 Smart Buildings

Smart buildings is the technology that automatically controls the various building operations. The discussion on this topic is beyond the scope of this chapter.

1.4.4 Virtual Machine Migration

The virtual machines (VM) can be migrated from one server to another server within data centers and outside data centers. The migration processes support operations like server consolidation, server maintenance, energy management, and traffic management. The VM's can be transferred in two modes: non-live migration and live migration. In non-live migrations, the virtualized machine is first suspended and necessary states are captured and transferred to the destination server. No communication with the virtual machine is possible during this time. In live migration, the processes are gradually shifted to the designated server after certain requirements are fulfilled. Three strategies are prevalent for live migration: pre-copy, post-copy, and hybrid-copy. There are overheads associated migration process are computation overheads, network overheads, and storage overheads. Much literature has been contributed toward the VM migration process. The future directions in this area are listed below [24, 25].

1.4.4.1 Optimized Process for Memory Migration

The VM's data in memory must be transferred to the designated machine to ensure that the process operations are not affected. Memory migration is one of the critical topics that can contribute toward optimizing the overall migration process. In pre-copy strategy, the process is transferred to the new machine after certain threshold conditions are fulfilled. It is felt that more optimized solutions are needed to ensure that termination of VM on the source machine is done safely. For post-copy and hybrid-copy there is a need to seek more robust solutions.

1.4.4.2 Migration of VM over Wide Area Networks

Migrating VM's in data centers connected through wide area networks (WAN) involves more complexities than migrating the VM's over LAN in the data center itself. Several factors like network bandwidth, data size, downtime of process, heterogeneity in architecture, and reliability need to be addressed.

1.4.4.3 Security

Transferring the VM and its associated data over WAN possesses a security risk. A strong cryptographic algorithm should be applied when transferring the data to safeguard from malicious activities. The bugs in software components that are part of the data migration process should be suitably patched to ensure they are not exploited during the transfer process.

1.4.5 Simulation Environments

Experimentation in the real environment is a complex problem. Furthermore, these experiments cannot be repeated since the many factors at the experiment time may change. The financial cost is another issue that an academic institution may not be able to bear. Various simulation frameworks have been proposed and widely used to evaluate the computer application. The interested reader can refer (Margariti, Dimakopoulos and Tsoumanis, 2020) (Fakhfakh, Kacem and Kacem, 2017) for further reading.

1.5   Emerging Trends

Based on fact-findings of cloud computing, this section focuses on recent research areas and usage of cloud applications in areas such as Wireless Sensor and Actuator Network (WSAN), Fog Computing, Internet of Things (IoTs), Vehicular Adhoc Network (VANET), Internet of Things, Sustainable Cloud Computing, Serverless Computing and Blockchains. In the Wireless Sensor Networks (WSN), motes or sensors monitor and record physical environment information and relay it to the computation's central location. The limitation in computing power, memory, energy consumption, and scalability drives the need for more resilient computing solutions. WSAN augments the WSN network by incorporating limited decision-making capabilities via actors or actuators. However, the massive amount of information collected from the physical systems and the limitation discussed above calls for the sensors cloud. As defined by IntelliSys "It is an infrastructure that permits the computations that is pervasive by using the sensors between the cyber and physical world as an interface, the cluster of data-computer as cyber backbone and Internet as the medium of communication" [28, 29]. The more emerging paradigms like Fog Computing and IoT have the research output fallen in this domain. Issues concerning storage of massive heterogeneous data, security problems including privacy and authentication, energy efficiency, network access management, and bandwidth management when cloud sensors increases need attention from the research community.

Fog computing is an emerging paradigm that brings the cloud closer to the devices that generate data. The huge amount of unstructured data that is produced needs to be segregated into times-sensitive data that requires immediate action by sensory application in the form of machine-to-machine communication or human- machine interaction. The less time-sensitive data can be pushed to the cloud for storage and analysis to gain insights into the system. Fog devices are located near nearby sensors and have computing, storage, and networking capabilities. These devices work collectively to minimize processing latency time by bringing compute closer to the data sources. These devices also save bandwidth by offloading the data that the device can process on the network edge. Some of the future directions in this computing field are standardization of SLA's for fog devices, Multi-Objective fog computing devices, mobile fog computing, green fog computing, SDN, and NFV support for fog computing and security issues [30].

Vehicular Adhoc Network (VANET) enables the vehicles to exchange the data for faster dessimation of information for the reliable and intelligent transportation system. Autonomous vehicles need to address the requirement of low-latency, high-mobility, scalability, real-time applications, and security. Fog computing, Software Defined Networks, and Cloud Computing can provide solutions for the present and future issues in VANET's.

Kevin Ashton coined the term Internet of Things (IoT). IoT is machine-to-machine interaction using the Internet. The technologies endorsed the concept of computing anywhere, anytime, using any device, in any location and any format or defined as a set of interrelated and interconnected systems that are collecting and transferring data through a wireless network without any human interruption. This transfer of data can be possible using embedded sensors. There are examples of IoTs like smart homes, smart cities, thermostats, and many more appliances. The devices, sometimes referred to as objects, should be identifiable, must access network and communicate, and able to sense, interact, and actuate with the environment. WSN, WSAN, Fog computing, and similar emerging paradigms is fuelling the growth of IoT. The sheer amount of information, bandwidth requirement, and computing resources integrate IoT with cloud computing. The cloud centric IoT approach addresses the problem of scalability, monitoring, deployment, and integration of devices for various applications.

The growth of the cloud computing industry has accelerated in recent years to fulfill the user's requirements. Many data centers have been set up worldwide to enhance computing power, increased storage, and better user data management. However, this also contributed to raised carbon footprints. One of the promising areas is *sustainable cloud computing* which caters to reducing environmental hazards. Usually, customers face the peculiar problem of "vendor lock-in" where they are locked in with one particular service provider. A new cloud strategy known as Omni cloud has been coined up where a customer can avail the services of more than one service provider. It is different from the hybrid cloud because various service providers jointly offer their service in collaboration with each other. Microsoft and Oracle have already linked their cloud services in the year 2019. Such a multi-environment cloud offers a variety of services to customers. The integration of technologies and applications of the multi-cloud environment is a challenging task.

Serverless Computing service is offered to supersede dockers, microservices, and Kubernetes to manage and deploy the application in the cloud environment. It facilitates the user to be more focused on developing and deploying the code. The developer buys the backend services on the "pay-as-you-go" model. The backend services are charged based on the computation instead of a fixed bandwidth or server. This feature enables the auto-scaling feature of the service. Serverless computing is offered in terms of database, storage, and Function-as-a-Service (FaaS). FaaS provides a cost-efficient method of implementing microservices. The idea is to execute the various piece of code on edge in response to an event. Some of the multiple challenges that need to be addressed are fault tolerance, security, and testing. The various challenges in the serverless can be referred to in [31].

Blockchain is distributed ledgers that enhance the security of the transaction in a decentralized environment. The blockchains can be successfully applied to various PaaS and SaaS applications to enhance security. Customers operating on the cloud platform can augment various transactions with blockchains. However, a robust infrastructure should be in place to supplement the resources required to compute and implement services of blockchains. Refer [32] for detailed issues like integerating blockchains with IoT, Transperency, Trustability vs. Privacy, and performance. Artificial Intelligence (AI) is an emerging paradigm that will fuel the future SaaS applications deployed on the cloud. The AI technology can optimize resource management, enhance security, and automize internal processes. Few AI-related services are IBM Watson, Microsoft Azure Machine Learning, AWS deep Learning, and Google Cloud Machine Learning Engines.

The other emerging topics to pursue research activities include security algorithms, interoperability between various heterogeneous devices, context-aware computing, SDN-NFV, and movement toward integrating various communication protocols.

1.6   Conclusion

Cloud Computing fulfills on-demand availability of computing resources, storage resources, network resources, middleware services, and application services. The cloud industry is expected to expand with a 17% compound annual growth rate to USD 832.1 billion by 2025. The current technological paradigms like IoT, Big Data, Fog Computing, Kubernetes, and Microservices are based on cloud-dependent solutions.

This chapter discussed basic terminologies and key terms associated with cloud-based computing. In spite of success and popularity, core issues like resources management, security, VM migration, and energy efficiency are still to be investigated. Optimal resource management will continue to be the key area for exploration. The inherent multi-tenant model and location independence will continue to raise privacy, confidentiality, and integrity issues. There has been an exponential increase in the incidents concerning cloud attacks. With the evolving technologies, security will be prominent research for years to come. Green computing practices and Virtual machine migration help manage issues like server consolidation, load balancing, and maintenance. We outlined various emerging trends in cloud computing. Future research needs to focus on the security aspect, heterogeneity of data, scalability, and energy-efficient system.

Bibliography

1 Mell, P. and Grance, T. (2012). *The NIST Definition of Cloud Computing: Recommendations of the National Institute of Standards and Technology*. Gaithersburg: NIST Special Publication 800–145. Computer Security Division, Information Technology Division, National Institute of Standard and Technology, 97–101.

2 Curran, K. *et al.* (2008). Autonomic Computing. In: Freir, M. M. and Periera, M. (eds) *Encyclopedia of Internet Technologies and Applications*. IGI Global, pp. 66–71. doi: 10.4018/978-1-59140-993-9.ch010.

3 Arockiam, L., Monikandan, S., and Parthasarathy, G. (2017). Cloud Computing: a Survey. *International Journal of Computer and Communication Technology* (2): 21–28. doi:10.47893/ijcct.2017.1393.

4 Broberg, J., Venugopal, S., and Buyya, R. (2008). Market-Oriented Grids and Utility Computing: The State-of-the-art and Future Directions. *Journal of Grid Computing* 6 (3): 255–276. doi:10.1007/s10723-007-9095-3.

5 Weiser, M. (2002). The computer for the 21st Century. *IEEE Pervasive Computing* 1 (1): 19–25. doi:10.1109/mprv.2002.993141.

6 Weiser, M. (1991). The Computer for the 21st Century. *Special Issue on Communications, Computers, and Networks*. September, 94–104.

7 David Linthicum. (2017). Chapter 1: Service Oriented Architecture (SOA). https://web.archive.org/web/20170707052149/https://msdn.microsoft.com/en-us/library/bb833022.aspx (accessed February 16, 2021).

8 Hugo Haas, W. and Allen Brown, M. (2004). Web Services Glossary. https://www.w3.org/TR/2004/NOTE-ws-gloss-20040211/#webservice (accessed February 16, 2021).

9 Mahmoud, Q. H. (2005). Service-Oriented Architecture (SOA) and Web Services: The Road to Enterprise Application Integration (EAI), Oracle. https://www.oracle.com/technical-resources/articles/javase/soa.html (accessed February 16, 2021).

10 Popek, G. J., and Goldberg, R. P. (1973). Formal requirements for virtualizable third generation architectures. *ACM SIGOPS Operating Systems Review* 7 (4): 121. doi:10.1145/957195.808061.

11 Terkaly, B., and Villalobos, R. (2013). Azure Insider—Microsoft Azure Web Sites: Quick-and-Easy Hosting as a Service | Microsoft Docs. *MSDN Magazine*, January. https://docs.microsoft.com/en-us/archive/msdn-magazine/2013/january/azure-insider-microsoft-azure-web-sites-quick-and-easy-hosting-as-a-service (accessed February 18, 2021).

12 Parikh, S. M., Patel, N. M., and Prajapati, H. B. (2017). Resource Management in Ccloud computing: Classification and Taxonomy. arXiv.

13 Mustafa, S. et al. (2015). Resource Management in Cloud Computing: Taxonomy, Prospects, and Challenges. *Computers and Electrical Engineering* 47: 186–203. doi:10.1016/j.compeleceng.2015.07.021.

14 Gonzalez, N. M., Carvalho, T. C. M. de B., and Miers, C. C. (2017). Cloud Resource Management: Towards Efficient Execution of Large-Scale Scientific Applications and Workflows on Complex Infrastructures. *Journal of Cloud Computing* 13. Springer Verlag. doi: 10.1186/s13677-017-0081-4.

15 Fard, M. V. et al. (2020). Resource Allocation Mechanisms in Cloud Computing: A Systematic Literature Review. *IET Software* 14 (6): 638–653. doi:10.1049/iet-sen.2019.0338.

16 Tabrizchi, H. and Kuchaki Rafsanjani, M. (2020). A Survey on Security Challenges in Cloud Computing: Issues, Threats, And Solutions. *Journal of Supercomputing*. Springer US. doi:10.1007/s11227-020-03213-1.

17 Mondal, A. et al. (2020). Cloud Computing Security Issues Challenges: A Review. *2020 International Conference on Computer Communication and Informatics, ICCCI 2020*, 20–24. doi:10.1109/ICCCI48352.2020.9104155.

18 Ahuja, S. P., and Muthiah, K. (2016). Survey of State-Of-Art in Green Cloud Computing. *International Journal of Green Computing* 7 (1): 25–36. doi:10.4018/ijgc.2016010102.

19 Ismail, L., and Materwala, H. (2018). EATSVM: Energy-Aware Task Scheduling on Cloud Virtual Machines. *Procedia Computer Science* 135: 248–258. doi: 10.1016/j.procs.2018.08.172.

20 Nguyen, Q.H. et al. (2013). A Genetic Algorithm for Power-Aware Virtual Machine Allocation in Private Cloud. *Lecture Notes in Computer Science (including subseries Lecture Notes in Artificial Intelligence and Lecture Notes in Bioinformatics)*, 183–191. Berlin, Heidelberg: Springer. doi: 10.1007/978-3-642-36818-9_19.

21 Ibrahim, A. et al. (2020). PAPSO: A Power-Aware VM Placement Technique Based on Particle Swarm Optimization. *IEEE Access* 8: 81747–81764. doi: 10.1109/ACCESS.2020.2990828.

22 Shrabanee, S. and Rath, A.K. (2020). SDN-Cloud: A Power Aware Resource Management System for Efficient Energy Optimization. *International Journal of Intelligent Unmanned Systems* 8 (4): 321–343. doi: 10.1108/IJIUS-07-2019-0032.

23 Potdar, A.M. et al. (2020). Performance Evaluation of Docker Container and Virtual Machine. *Procedia Computer Science* 171 (2019): 1419–1428. doi: 10.1016/j. procs.2020.04.152.

24 Zhang, F. et al. (2018). A Survey on Virtual Machine Migration: Challenges, Techniques, and Open Issues. *IEEE Communications Surveys and Tutorials* 20 (2): 1206–1243. doi: 10.1109/COMST.2018.2794881.

25 Noshy, M., Ibrahim, A., and Ali, H.A. (2018). Optimization of Live Virtual Machine Migration in Cloud Computing: A Survey and Future Directions. *Journal of Network and Computer Applications* 110: 1–10. doi: 10.1016/j.jnca.2018.03.002.

26 Margariti, S.V., Dimakopoulos, V.V., and Tsoumanis, G. (2020). Modeling and Simulation Tools for Fog Computing - A Comprehensive Survey from A Cost Perspective. *Future Internet* 12: 5. doi: 10.3390/FI12050089.

27 Fakhfakh, F., Kacem, H.H., and Kacem, A.H. (2017). Simulation Tools for Cloud Computing: A Survey and Comparative Study. *Proceedings – 16th IEEE/ACIS International Conference on Computer and Information Science, ICIS 2017*, 221–226. doi: 10.1109/ICIS.2017.7959997.

28 Tan, K.-L. (2010). What's NExT? *Proceedings of the Seventh International Workshop on Data Management for Sensor Networks – DMSN '10*, 1. New York, New York, USA: ACM Press. doi: 10.1145/1858158.1858160.

29 Alamri, A. et al. (2013). A Survey on Sensor-Cloud: Architecture, Applications, and Approaches. *International Journal of Distributed Sensor Networks* 2013. doi: 10.1155/2013/917923.

30 Yousefpour, A. et al. (2019). All one needs to know about fog computing and related edge computing paradigms: a complete survey. *Journal of Systems Architecture* 98 (February): 289–330. doi: 10.1016/j.sysarc.2019.02.009.

31 Hendrickson, S. et al. (2016). Serverless computation with OpenLambda. *8th USENIX Workshop on Hot Topics in Cloud* Computing, *HotCloud 2016*, 33–39. USENIX Association. https://dl.acm.org/doi/10.5555/3027041.3027047 (accessed May 16, 2021).

32 Song, J. et al. (2021). Research Advances on Blockchain-As-A-Service: Architectures, Applications and Challenges. *Digital Communications and Networks*. doi: 10.1016/j. dcan.2021.02.001.

2

Cloud IoT

An Emerging Computing Paradigm for Smart World

Ruchi Bhatnagar, Prof (Dr.) Paramjeet Rawat and Dr. Amit Garg

School of Computer Science and Applications, IIMT University, Meerut

2.1   Introduction

The IoT has revolutionized and surprised the world by widespread practical applications in multiple fields, such as wearables to health monitoring, traffic management to fleet monitoring, smart homes to smart cities, smart agricultures to manufacturing and industrial plants, smart grids to maintenance management, etc. [1]. It is an emerging model that provides connectivity, intelligence, and identity with scalability the Internet in order to facilitate our lives [2]. It empowers collaboration from anywhere; becomes an important aspect of life that can support and facilitates devices and users. IoT is an invention that puts together wide variety of technologies, frameworks, intelligent devices sensors, and IPv6. Moreover, it takes advantage of data processing at network edges or remote server, quantum and nanotechnology in terms of storage, green computing in terms of environment, processing speed in terms of 5G; which were not conceivable earlier [3]. The IoT is not only dedicated to connect remote devices to achieve flawless functioning and simplify operations but also act as enabler and integrators to build applications, collect data remotely, connect securely, and manage devices. It also helps to overcome many challenges, through increased capacity, greater intelligence, and real-time operational insight to shaping the future world.

For the intelligence and interconnection, IoT is not referred as single technology; rather it is a stack of various technologies and brings competitive IoT devices closer that work together in conjunction. Deploying and managing applications in an IoT scenario can be challenging, costly, time consuming, and require strategic planning especially when computational offloading and security is a concern. The cloud paradigm enables you to increase your margins, provide quickly and faultlessly process of voluminous data, and outperform the competition based scenario and reduce downtime while rapidly deploy applications. Thus, the incorporation of IoT and cloud computing offers data communication, application management, and optimization as more and more devices are coming closer to make a system that drastically changes the potentials of gadgets. The underlying idea behind IoT and the cloud computing is increase efficiency in the day-to-day tasks by ensuring easier exchange of data between IoT devices based on multiple platforms. Since the collaboration relationship is joint, both the services complement each other efficiently. The IoT

Emerging Computing Paradigms: Principles, Advances and Applications, First Edition.
Edited by Umang Singh, San Murugesan and Ashish Seth.
© 2022 John Wiley & Sons Ltd. Published 2022 by John Wiley & Sons Ltd.

becomes the source of the services, while the cloud becomes the ultimate destination for underlying infrastructure, servers, and storage. Cloud computing is on demand availability of computing resources via the Internet, such as software, storage, and even infrastructure. The influx of this technology has created a worldwide competitive market, its economical perspectives and increased productivity possibilities are enormous [4]. In cloud computing resources are connected via broadband connections, with the goal of maximizing computing and minimizing cost. Resources accessed by enabling users and download the data on chosen devices as opposed to being physically present. Cloud services are characterized by five key concepts such as on-demand service, broad network access, resource pooling, rapid elasticity, and measured service. As smart world is shaping, a lot of changes are happening; in cloud IoT paradigm; some of these changes will be adaptive and adjustable to meet the ongoing demand of future. Considering all the aspects of smart environment with the cloud computing model the continuous improvement has been desirable for shaping new world. The handshaking of IoT with cloud computing is not a new coin for industries, but needs an ever changing upgradation to meet the successful needs of customers.

This chapter illustrates the brief advancement of cloud computing, cloud IoT recent integration development trends, related issues, and analytical application areas with Cloud IoT infrastructure.

The objectives of this chapter focus on:

1) An integration view of cloud IoT with advancement.
2) Summary of key cloud IoT issues.
3) Particularization of cloud IoT integrated application areas.
4) Conclusion with future research directions for researchers and practitioners.

The remaining part of the chapter has been organized as follows: segment 2 reviews advanced concepts of cloud computing; segment 3 elaborates IoT and cloud integration paradigm; segment 4 defines advanced cloud IoT integrated technologies, segment 5 briefs related issues faced by IoT-Cloud paradigm; segment 6 elaborates smart application areas; and segment 7 concludes the chapter.

2.2 Cloud Computing

Automation and application are the future proof for IT world. The cloud, with its design, allows powerful and expensive systems, and roll out an entire corporate landscape with no human interaction [5]. It has an enormous growth to deploy and maintain software and is being meet infinite demands of industries widely. It defines a paradigm, in which real-time scalable resources and third-party services, allows access over the Internet rather than having local servers or personal devices. It is the technological model; to use IT infrastructures and provide services by the help of apps and becomes an appropriate resource [6]. Even though there are numerous variations on the functions of cloud computing; it provides technological capabilities such as act tactical enabler, provide simplified management, align loads, broad Network Access, Resource Pooling, and innovative acceleration.

Cloud computing is a model that relies on shipping high volumes and has a very simple concept that can leverage competitive advantage in the IoT. A simple explanation for cloud computing is delivery of different services such as servers, storage, networking, databases, software analytics, intelligence using virtual pool of resources via the Internet, and is available anywhere via net-enabled services. The essence of cloud computing is utility computing. The information being accessed is found in the "cloud" and covers whole range of applications, from monitoring, metering, sensing, to actuating in an IoT scenario. Companies are not required to own their own servers and can use capacity leased from third parties; in addition, executed carefully designed automation strategies and energy-saving plans. It enables to create future-ready innovations across all of leased environments using the newest technologies. The recent sparks in cloud collaboration with IoT solution not only focused on telemetry and data collection from devices but also to fulfill complex scenario, for supporting the business needs. Its new infrastructure is upgrading and comprises many cloud components in terms of automation, smart service design, software changes adoptability, easy to operate, etc. engineered to leverage the power of cloud resources to make the environment resilient to solve business needs constraints. In a traditional design environment, you'll connect your database to modules, and those modules will connect with an app but as new demands are approaching the environment at a high level, new cloud-native architecture needs; that allows adaptive and agile development of new services with very different set of architectural constraints offered by the cloud compared to traditional application and services. It defines the components as well as the relationships between them and should be largely self-healing, cost efficient, dynamically orchestrated and easily updated and maintained through service-oriented and event-driven strategies.

2.2.1 Cloud Evolution

In the last two decades the cloud and accommodating models have experienced major transformation from cloud hosted to micro services and serverless architecture. The deep insight about these transformations is as follows:

2.2.2 Monolithic Cloud Hosting Architecture

The services of cloud computing hosting started by the first decade of the 21st century. These three tier application use cases require abilities like distributed computing, network, storage, and computation, etc. The consumers were taking advantage of elasticity of the resources to scale up and down based on the demand. The adaptable functionality introduced in Infrastructure as a Service (IaaS) and Platform as a Service (PaaS) allow for a single instance for the sake of scalability. However, the architecture means that it's all in one competences cannot pay for the replication of instances for other purposes, such as having multiple versions, or as a by-product of deployments. It offered three different functional areas illustrated in Figure 2.1.

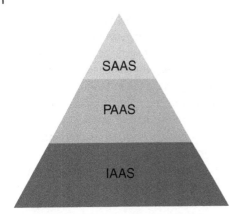

Figure 2.1 Cloud hosting services.

Infrastructure as a Sservice (IaaS): This one is most malleable from all three flavors of services, but act as a pay-as-per model; allows to outsource their IT infrastructure such as hardware, OS, and networking.

Platform as a Service (PaaS): This service of cloud computing offers platform to their clients for offering to run different business applications; thus the cloud developers does not need to worry about all the server infrastructure, network, and monitoring tasks.

Software as a Service (SaaS): This is a software distribution model; allows users to connect and use software applications as a whole without responsibility even for the application code. This offers a variety of benefit subscription options that provide the software services out of the box but it's inflexible if the client needs to have any custom business functions outside of what's offered by the provider.

2.2.3 Micro Service Cloud Architecture

It is an architectural concept with cluster of micro services that differentiate the monolithic paradigm into smaller independent units. They are powerful, and enable smaller teams to own the full-cycle development of specific business and technical capabilities. Developers can deploy or upgrade code at any time without adversely impacting the other parts of the systems (client applications or other services). The services can also be scaled up or down based on demand, at the individual service level. A client application that needs to use a specific business function calls the appropriate micro service without requiring the developers to code the solution from scratch or to package the solution as library in the application. The micro services approach encouraged a contract-driven development between service providers and service consumers. This sped up the overall time of development and reduced dependency among teams. In other words, micro services made the teams more loosely coupled and accelerated the development of solutions, which are critical for organizations, especially the business start-ups.

2.2.4 Serverless Cloud Architecture

A hot trend that revolutionized the world and gained a lot of attention in the last few years is serverless architecture, also recognized as serverless computing. Serverless computing is a step further than the PaaS model in that it scales automatically and fully abstracts from the application developers. In this computing platform, an illusion has been created, originally for the benefits of developers whose business services hosted at cloud; and extended the way people use it.

These backend services provide clients a platform to write and deploy codes without the worrying of underlying infrastructure; and to pay on the basis of computation done instead of fixed charge. The backbone of this computing arena is lower cost, simplified scalability and quicker turnaround time; while the cold start of same requested functions becomes necessary trade-off of Function-as-a-Service (FaaS).

2.2.4.1 Comparative Statistics

The investigation study based upon cloud application functionality and some nonfunctional requirements, an appropriate monolith, micro services, or serverless model selects for each specific use case. These all are known as solution architecture for clients as per their needs. The comparison statistics of these three depicted in Table 2.1.

While choosing any of appropriate cloud paradigm; the advanced cloud architecture empowers to build and run scalable applications to meet the vision of cloud applications, encouraging the diverse applications facilities; their delivery and integration with IoT based use cases.

Table 2.1 Comparative statistics.

Architecture Type	Key Feature	Functional Services	Drawbacks
Monolithic Cloud Hosting Architecture	Easy to develop, deploy, and scale	Application that has different modules, where no single developer understands entirely an application; examples ERP, CRM	High maintenance,difficulty to adopting new and advanced technologies
Micro Service Cloud Architecture	Better deployment ability, relatively small and productive, improved fault isolation	Application modules are independent of each other thus known independent deployment object's and can be scaled independently of other services. examples: customer service order service and inventory service	Additional complexity, testing is more difficult, increased memory consumption
Serverless Cloud Architecture	Automatic scalability, quick start, pay-as- you-go model	Application modules can be broken down into single functions, and provides actionable results. Examples: Authentication, notification, event streaming	Shut down when not in use, every new user demanding same functions again needs cold start

2.3 Cloud IoT

The dynamic and global network IoT and on-demand servicing of cloud computing represents bilateral relationship, benefitting from innovation and giving a competitive edge. With connectivity becoming increasingly necessary in our everyday working and social lives. The countless application of IoT is encompassing Internet connectivity among billions of devices and produce a huge amount of data. The production of such big data storage and transformation demand intelligent network control and management solutions while the activities like storage and computation take place in the cloud platform rather than on the device itself. Thus, millions of globally dispersed devices within IoT infrastructure connect and managed and ingest data using cloud services.

It is in combination with other services on a cloud IoT platform, that provides a complete solution for collecting, processing, analyzing, and visualizing IoT data in real time to support improved operational efficiency of the network. Using the cloud also allows for high scalability. When you have hundreds, thousands, or even millions of sensors, putting large amounts of computational power on each sensor would be extremely expensive and energy-intensive. Instead, data can be passed to the cloud from all these sensors and processed there in aggregate. For most of the IoT applications, the master brain of the system is in the cloud; sensors and devices collect data and perform actions, but the data processing, intelligent commanding, and smart data analytics typically happens in the cloud. Nowadays, data, as well as IoT devices, grows exponentially and the data processing and commanding could take place locally rather than in the cloud via an Internet connection; known as "fog computing" or "edge computing," which makes a lot of sense for energy constraint sensor-based IoT applications and make enabling computer paradigm for smart world. These edge networking and IoT have had a profoundly positive impact on world, however, their recent convergence has surfaced interoperability challenges across platforms, applications, and systems.

Cloud computing and the IoT are two trend technology paradigms of recent time and also two main enablers for smart manufacturing and digitization. The relationship between IoT and cloud computing creates ample opportunity for business to harness exponential growth. Put simply, IoT is the source of data and cloud computing is the location for storage, scale, and speed of access. Smart products can have advantages over their standard variants and customers begin to expect certain levels of monitoring, smartness, and scalability making these technologies more and more a necessity for many enterprises [7]. IoT and cloud computing also complement one another and both working together to provide an overall better IoT service. However, there are crucial differences between them, making each of them an effective technical solution separately, as well as together. The role of cloud computing in IoT works as part of a collaboration and is used to store IoT data. The cloud is a centralized server containing computer resources that can be accessed whenever required. Cloud computing is an easy method of travel for the large data packages generated by the IoT through the Internet. These both in collaboration increasing the efficiency of everyday tasks. While IoT has penetrated mainstream technology and market place, it generates a massive amount of big data. Besides, cloud computing paves the way for this enormous data. From a storage solution to accessing data remotely, IoT and cloud computing builds an integration. Not only storage and access, but there are also many areas where we can do a fit-gap analysis between IoT and cloud computing.

2.3.1 How Does It Work?

Cloud computing and IoT are two enabling technologies often paired together, but it is also necessary to know that how exactly do the two entities interact with each other? While the IoT connectivity could exist without the cloud, it's safe to say that cloud computing enables many IoT devices to function with much greater power and efficiency. In an IoT network, many IoT sensors (dozens, hundreds, or thousands) collect data and send it to a central location for analysis. This data is important to be analyzed in real time with proper analytics tools so that the faults and failures can be resolved in minimal time, which is the core purpose of this integration. Cloud computing helps by storing all the data from thousands of sensors (IoT) and applying the needed rule engines and analytics algorithms to provide the expected outcomes of those data points. The cloud providers also allow smart companies functionality to store and process copious amounts of data at minimal costs, opening the door to big data analytics. Without the cloud, aggregating IoT data across large areas and different devices is far more complicated. Cloud IoT includes the underlying infrastructure, servers, and storage, needed for real-time operations and processing with services and standards necessary for connecting, managing, and securing different IoT devices and business applications. The integrated cloud IoT architecture is depicted in Figure 2.2.

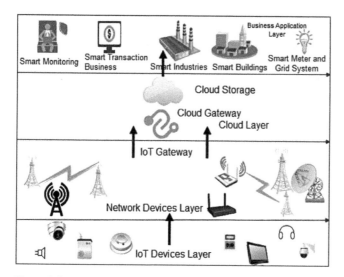

Figure 2.2 Cloud IoT integrated architecture.

For example, a smart city can benefit from the cloud-based deployment of its IoT systems and applications. A city is likely to deploy many IoT applications, such as applications for smart energy management, smart water management, smart transport management, urban mobility of the citizens, and more. These applications comprise multiple sensors and devices, along with computational components. Furthermore, they are likely to produce very large data volumes. Cloud integration enables the city to host these data and applications in a cost-effective way. Furthermore, the elasticity of the cloud can directly support expansions to these applications, but also the rapid deployment of new ones without major concerns about the provisioning of the required cloud computing resources creates some issues and challenges for smart platform.

2.4 Emerging Concepts of Cloud IoT

The IoT is not a concept but a true technological network of all networks around the world, being able to connect with anything, be it smartphones, vehicles, people, or even animals by identifying them with unique identifiers (UIDs). It is increasing and growing everyday with better ideas because of new advancements in technologies and concepts like smart homes, thermostats, and sensors. Tracing and monitoring different objects is now required for security and comfort. When it met with the cloud, new era of custom tailoring applications led to the development of an extremely profitable deal for people in the near future technologies. Some of its emerging concepts boots up its collaboration and provides numerous resources-based services as defined below.

2.4.1 Edge Computing

The three main gears of the IoT are physical things, communication networks and the cloud. In countless current IoT systems, the actual things have limited computing power and its main purpose is to transmit sensor data over communication networks to a cloud backend. With minimum local data processing involved and the basic approach of performing all processing in the cloud is not scalable and finds its limits as the number of devices sending data grows, or data volume from a device exceeds a certain amount. So that a distributed computing framework called edge computing introduces that brings applications closer to data sources as IoT devices or local edge servers. This belonging of data at edge servers deliver many enterprise benefits such as faster accesses, improved response time, and improved bandwidth availability. This intelligent integration enabled IoT to store, process, and analyze data on the edge of the network, with increasing edge intelligence and connectivity that drives intelligent systems. It states that edge intelligence will be the key value that delivers exceptional growth for IoT implementation. This integrated computing trend by IoT sensor nodes, edge computing, software, and cloud services for a wide array of IoT applications. Edge computing in the context of IoT means that data is processed at the place where it is generated. Pre-processing, filtering, analytics, and even artificial intelligence (AI) are done on the edge device instead of central point. This also helps to reduce latency, data volumes, and loads on cloud infrastructure, while at the same time improving responsiveness, robustness, and resiliency of applications. Edge computing can be done on advanced Programmable Logic Controllers (PLCs), Programmable Automation Controllers (PACs), Industrial PCs, or IoT gateway devices in industrial IoT applications. Futuristic advancement in networking technologies, such as 5G wireless, are allowing for edge computing systems to accelerate the creation or support of real-time applications, such as video processing and analytics, self-driving cars, AI, and robotics.

2.4.2 Fog Computing

Another advanced cloud IoT integration paradigm is fog computing; it can be seen as an extension of edge computing where computing, storage, control, and networking functions are distributed locally and routed to the cloud using Internet backbone. Fog computing means additional computing and networking layers between edge and the cloud. In nature, fog exists between ground and clouds, same fog computing sits between physical things and cloud computing. An interesting and important aspect of fog computing is that it takes the concepts from cloud computing and brings them closer to the things network. The fog extends the cloud services to be closer to the things network that produce and act on IoT data. These special devices, called fog nodes, can be deployed anywhere with a network connection: for example on a factory floor, on top of a power pole, alongside a railway track, in a vehicle, or on an oil rig. Any device with computing, storage, and network connectivity capability can be a fog node. Examples include industrial controllers, switches, routers, embedded servers, and video surveillance cameras. Fog nodes considered by analyzing the most time-sensitive data at the network edge, close to where it is generated instead of sending vast amounts of IoT data to the cloud, perform action on IoT data in milliseconds, based on policy and sends selected data to the cloud for historical analysis and longer-term storage. Fog computing is not a replacement of cloud computing by any measure, it works in conjunction with cloud computing, optimizing the use of available resources creating data-path hierarchy and transmitting only summary and exception data to the cloud that leads to increased business agility, improved security, and higher quality of service levels.

2.4.3 Mist Computing

While fog computing is the answer to many challenges in the cloud IoT domain, such as high bandwidth requirements and manageability of applications, this concept can be further extended by pushing the computation to the end devices. In fog computing the gateway bears the responsibility for IoT application execution, regardless whether the application is just simple data collection or building automation with many actuation tasks. However, this approach can also have some drawbacks, such as increased delays in applications involving control, unnecessary high bandwidth requirements as all data must move through the gateway. The gateway is a single point of failure for applications that must be executed on the network and the operation of the entire network is dependent on the gateway; then a new concept of mist computing introduces.

Mist computing takes fog computing concepts even further by pushing appropriate computation to the very edge of the network, to the sensor and actuator devices that make up the network. With mist computing the computation is performed at the edge of the network in the microcontrollers in the embedded nodes. The mist computing paradigm decreases latency and further increases the autonomy of a solution. By applying the principles of service-based architecture among the end devices the application can be described as a combination of services, which are dependent on each other. Any device that has access to the network can subscribe to a service that is offered by any of the devices on the network. It is utilized at the extreme edge of a network that consists of microcontrollers and sensors. By working at the extreme edge, mist computing can harvest resources with the help of computation and communication capabilities available on the sensor. Mist computing infrastructure uses microcontrollers and microcomputers to transfer data to fog computing nodes and eventually to the cloud. Using this network infrastructure, arbitrary computations can be processed and managed on the sensor itself. Each device in the network must be aware of its location as most applications tend to be location dependent. The necessary "location awareness" can be created at installation time (by "telling" the device, what its location), or the devices can determine their location autonomously by determining their location relative to some existing beacons, with known locations. Applications will be able to support this relative location functionality and incorporate this into mist. The services provided by end devices may also be requested by mobile devices or servers, in which the service request reaching a specific network is routed to the device, that is able to provide the specific service. This means that in one network end devices and a gateway may be both providing services to the same server. As an example in a building automation scenario we may be interested in room occupancy information for every single room, so all the occupancy sensors in the individual rooms must report information directly to the server, while the operation times of standalone air conditioning (AC) units may be aggregated (in the gateway) to estimate the total power usage of AC units in the building.

2.4.4 Cloudlets

The mobile as well as other IT devices nowadays are being developed embedded with a number of advanced features such as augmented reality, face recognition, natural language processing, gaming, video processing, 3D modeling software, etc. These applications usually are resource-hungry, requiring intensive computation and high energy usage. But the mobile devices are resource constrained in terms of processing power and battery life. So, in order to execute these types of applications, the resource intensive applications are uploaded to the cloud using a mechanism called offloading where all these processing can be carried out in cloud using the resources there, and the results are send back to the IT devices in our hand. Based on the type of tasks and the needed resources, the whole process or a part of the process get offloaded to the cloud for processing. But as I mentioned above in the edge computing section, sending data from data resources to clouds that are miles away have latency and bandwidth issues. And, if there is a situation where the Internet service provider failed to conserve the connection between the device and the cloud server, there will be delays, packet loss, and interrupt user experience. So, in order to avoid and reduce these problems, the cloudlet concept was introduced in a cloud computing paradigm.

A standard definition for cloudlet is "cloudlets are mobility-enhanced small-scale cloud data centers that are located at the edge of the Internet." So, by using cloudlets, the resource intensive tasks can be offloaded to it for processing hence will reduce latency, bandwidth, and save a lot of time. Cloudlets' latency and bandwidth advantages are especially relevant in the context of automobiles, to complement vehicle-to-vehicle approaches being explored for real-time control and accident avoidance. During failures, a cloudlet can serve as a proxy for the cloud and perform its critical services. Upon repair of the failure, actions that were tentatively committed to the cloudlet might need to be propagated to the cloud for reconciliation. Including these privacy and security conservation are other benefits of cloudlets. While using cloud for processing, our secure data have to travel to cloud servers' miles away, hence security of the data will be in sometimes breached. Hence, by using cloudlets, all the private data will be processed at the edge of devices and help in the conservation of the security and privacy of data [8–10]. Enterprise users must understand that the edge, fog, mist, and cloudlets computing paradigm have their own set of strengths and weakness their conceptual architecture referenced in Figure 2.3.

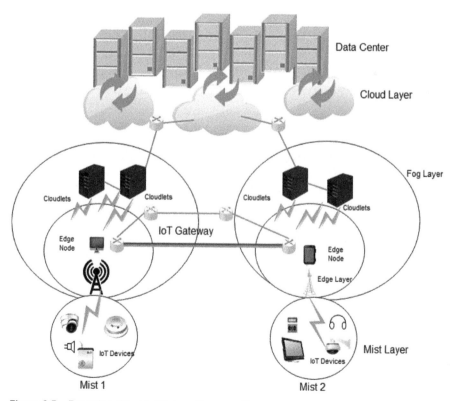

Figure 2.3 Emerging cloud-IoT integration paradigm.

Table 2.2 Summarization details of emerging cloud-IoT technologies.

	Edge Computing	Fog Computing	Mist Computing	Cloudlets
Devices	Devices where sensors are attached	Routers, switches, access points, gateways	Microcontroller at devices	Data center in a box
Data Processing	Directly on devices where attached	Cloud or at local data center	Extreme edge of network that consists of a microcontroller or sensor	Resource rich computer/ cluster
Software Architecture	Layer 2 edge/ connectivity layer	Fog abstraction layer	Beneath fog abstraction	Cloudlet agent based layer
Context Awareness	Low	Medium	Low	Low
Accessing Mechanism	Internet	Bluetooth, Wi-Fi, mobile networks	Mobile network	Wi-Fi
Specific Application areas	connected homes to perform tasks like turning on the heater or lights in near real time	In smart cities, where many devices use real-time data to perform various tasks	Incredibly useful for IoT in public transportation as the devices may not be stationary and may only serve a singular purpose	Supporting resource-intensive and interactive mobile applications by providing powerful computing resources to mobile devices with lower latency

Understanding and using these paradigms correctly will help in ensuring that the growing number of IoT devices can work efficiently. Also, businesses can utilize fog, mist, edge, and cloudlets computing together to utilize their strengths and minimize their limitations. As these networking architectures complement each other, businesses can use them to design secure, reliable, and highly functional IoT solutions for smart world yet their comparative analysis also requires to choose any one of them or in the form of integration as based upon IoT application use case. The summarization details of these presented in Table 2.2.

2.5 Advanced Infrastructure of Cloud IoT

The cloud IoT platform must monitor IoT endpoints and event streams, analyze data at the edge and in the cloud, and enable application development and deployment. A smart IoT infrastructure needs, digitization and an interconnected collection of networks that transmit data; and offers economics to applications. Number of advanced cloud-IoT collaboration schemes are in charge of all communications across devices, networks, and cloud services that make up the IoT infrastructure practically feasible as per need of various application scenarios. This chapter elaborates four architecture choices for enabling digital transformation of IoT enabling services, depending on needs.

2.5.1 Monolithic Cloud-IoT Architecture

The oldest and traditional way of integrating applications using monolithic approach comprises a client-side user interface, a server-side application, and a database. It is unified and all the functions are managed and served in one place. It is one large code base and lack of modularity leads to issues that if developers want to update or change something, they access the same code base. So, they make changes in the whole stack at once. In spite of having functionality such as less crosscutting concerns, easier debugging and testing, simple to deploy, and development becomes cumbersome with growing IoT devices and real-time versioning of updates with hard management, not scalable, and sometimes act as a barrier for new technologies. It was designed for application development before the proliferation of public cloud and mobile applications and has difficulty to adapting an application that become too large and complex to make frequent changes. Not only that, but it also requires the maintenance of at least three layers of hardware and software, that can make it inefficient (Figure 2.4).

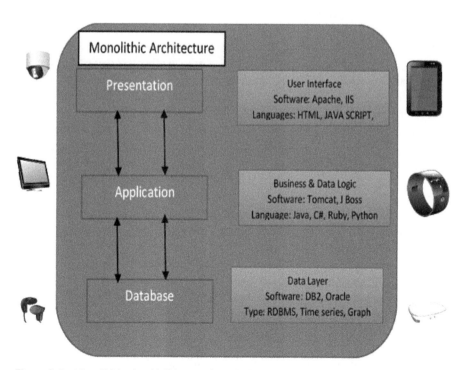

Figure 2.4 Monolithic cloud IoT integration platform.

2.5.2 Micro Service Cloud-IoT Architecture

Independent components, easier understanding, better scalability, flexibility, and higher level of agility advances the concept of micro service cloud-IoT integration. These independent units carry out every application process as a separate service. So all the services have their own logic and the database as well as perform the specific functions. As each service covers its own scope and can be updated, deployed, and scaled independently the extra complexity, system distribution, cross-cutting concerns, and multitude of independently deployable components testing becomes much harder in this model. While the migrating to micro services cloud-IoT integration allows smooth adding of new SaaS services: anomaly detection, delivery prediction, route recommendations, object detection in logistics, natural language processing for document verification, data mining, and sensor data processing. One of the best choices for creating and running micro services application architectures for IoT is by using containers; that encapsulate a lightweight virtualization runtime environment for your IoT application and allow you to move the application from the developer's desktop all the way to production deployment. You can run containers on virtual or physical machines in the majority of available operating systems. Containers present a consistent software environment, and you can encapsulate all dependencies of your application as a deployable unit. Containers can run on a laptop, bare metal server, or in a public cloud (see Figure 2.5).

2.5.3 Serverless Cloud Integration

An architectural integration with IoT is handling your cloud environment's underlying servers so that you don't have to. It allows developers to focus on writing code without needing to worry about deploying, managing, and scaling servers. PaaS, serverless computing allows IoT businesses to offload all of a server's typical operational backend responsibilities, freeing their developers, and engineers to work on creating new products and services. This service's architecture solves insights from your global device network with an intelligent IoT platform whose scalable, fully managed integration lets you connect, store, and analyze data at the edge and in the cloud. It is container-based environments; empower to run applications in public, private, and hybrid clouds. It supports people, processes, and technologies to build, deploy, and manage apps that are ready for the cloud. Serverless IoT architecture also help to provide IoT businesses with incredible savings, scalability, and performance. It's working based on decoupled systems that run in response to actions. This event-driven architecture uses events to trigger and communicate between decoupled services. Electronic Design Automation (EDA) has been here for a long time, but it now has more relevance in the cloud. For the serverless model, there is no server management needed. The serverless model is also quickly scalable (so quick updates and deployment are possible) and it is stateless (see Figure 2.6).

These advanced collaborative platforms facilitates new emerging applications and requires following key principles to follow best practices and work well with integrated platforms and application based use cases of IoT.

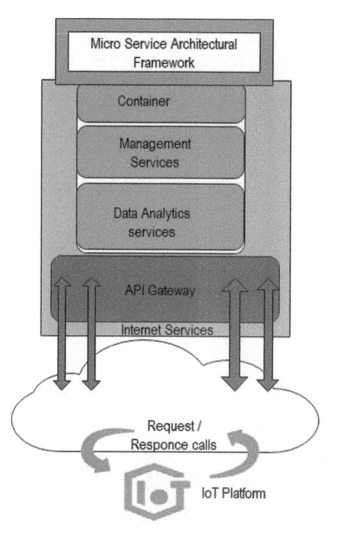

Figure 2.5 Micro service cloud-IoT integration.

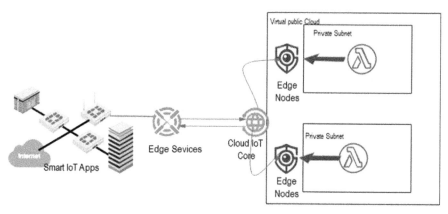

Figure 2.6 Server less cloud IoT platform.

2.5.3.1 Resiliency

An evolved cloud models needs deployment across multiple physical data centers, using multiple decoupled tiers of the application, and automating the start-up, shutdown, and migration of application components between cloud locations to ensure resiliency for the application.

2.5.3.2 Semantic Versioning & Parallelization

Designing an application that can execute distinct processes in parallel with other parts of the application will directly impact its ability to have the performance required as it scales up; thus evolving model of cloud platform allowing the same set of functions to execute many times in parallel, or having many distinct functions in the application execute in parallel.

2.5.3.3 Event Driven

Applications that are event-driven are logged and analyzed by advanced machine-learning techniques to enable additional automation to be employed; becomes a key principle requirement of new scenario cloud platform.

2.5.3.4 Security

A secure cloud environment is back bone of all applications to ensure and takes advantages of secure cloud services, design level securities in all layers as well as reduces the blast radius in scale of event failure and breach.

2.5.3.5 Future Proofing

Cloud plays an important role to ensure that an application will continue to evolve along the platform as time and innovation moves on. Implementing key principles will help with future proofing; however, all applications must be optimized through automation and code enhancements constantly to always be able to deliver the best cloud platform results and make the computing more and more pervasive.

2.6 Cloud-IoT Issues

The IoT and cloud computing both are latest technology buzz and megatrend of the computing industry. Cloud computing empowers an appropriate, on-demand, and scalable network access to a shared pool of configurable computing resources; while the IoT is a smart technology that helps all networked connected devices to update themselves according to changes in the surrounding environment and to be able to be adopted and work in any other strange environment with high accuracy. The cloud and IoT integration allows new scenarios, for smart services and applications, as SaaS, Data Base as a Service (DBaaS), Video Surveillance as a Service (VSaaS), and many more. As every technology comes with a baggage of some pros and cons; similarly, cloud-IoT integration too comes with its share of issues despite being core strength. This integration also creates some major problems under some rare circumstances. IoT and cloud integration involves several challenges and issues as services interoperability, power and energy efficiency of devices for data transmission and processing, big data generated by several devices, security and privacy, integration methodology, network communications, storage, etc.

2.6.1   Interoperability Issues

Interoperability refers to the basic ability of computerized systems to connect and communicate with one another readily, even if they were developed by different manufacturers in different environment. Being able to exchange information between applications, databases, and other computer systems is crucial for the modern world. The interoperable environment of heterogeneous IoT devices and cloud services is an important issue in a cloud-IoT integrated environment. The key concerning issues of interoperability are technical, syntactic, semantic, and organizational interoperability; that requires a closer look at interactions of the key components in IoT ecosystems regarding different aspects [11].

2.6.2   Energy Efficiency

The integration of the cloud and the IoT is attracting attention to the industry. Particularly, the data (alarm, security, climate, and entertainment) gathered by sensors are transmitted first to the gateway, that then transmits the received sensory data to the cloud that quickly consumes the node energy. Eventually, the cloud stores, analyzes, processes, and transmits the sensed data to the users on demand. During the entire data transmission process, if the data transmission from the sensor nodes to the cloud is not succeeded, data are retransmitted until they are successfully delivered, requires efficient energy mechanism for collaborated environment, and remains a significant open issue. The reliable delivery of data in cloud-based IoT is a big concern for a cloud-IoT environment because of limited power constraint of sensor nodes in IoT [12].

2.6.3   Big Data Generation and Processing

With the evolvement and development of smart home, smart cities, and smart industries; the cloud paradigm empower them to collect huge data by heterogeneous connected devices and exchange to fulfill user needs. As this collaboration technique benefitted users in so many ways some concerning issues include big data is never 100 percent accurate, it's important to be sure before analyzing data that the sensors function accurately and the quality of the data coming for analysis is reliable and not spoiled with factors such as breakdowns in sensors. The physical layer of paradigm containing connected things generate terabytes of data, and it's a demanding task to choose which data to store and which to drop while on which data needs quick analysis and which requires deeper analysis needing the advanced cloud models such as edge data analytics, cloud gateway and machine learning modules.

2.6.4   Security and Privacy

The cloud and IoT that refers to the integration of the cloud computing and the IoT, has dramatically changed the way treatments are done in the ubiquitous computing world. This evolving integration has become vital because the important amount of data generated by IoT devices needs the cloud model as a storage and processing infrastructure. The security issues in cloud IoT stay more critical since users and IoT devices continue to share computing as well as networking resources remotely. Among the main concerns of cloud IoT, privacy and data integrity have a great place. The one of main constraint of the exchanged data is sharing of the user's personal information and keep it safe. Several issues such as the protection of user's privacy and manufacturer's IP; the detection of malicious activity and how to block them, come under cloud and IoT security threats [13] as it continuous growth. One particularly important issue that has not yet been solved is how to provide appropriate authorization policies and rules while ensuring that only authorized users have access to the sensitive data.

2.6.5 Integration Methodology

Cloud computing has resolved most of IoT issues. Truly, the IoT and cloud are two comparatively challenging technologies, and are being combined in order to change the current and future environment of Internet-working services [14]. This needed a standardization mechanism, standard protocol formats, high bandwidth, compatible architecture, and standard application programming interfaces to allow for interconnection between heterogeneous smart things and the generation of new services, which make up the successful integration of cloud-based IoT paradigm.

2.7 Application Areas

An IoT enables an innumerable benefits of different business applications and promises to bring immense value into our lives with superior sensors, excellent technologies, and revolutionary computing capabilities. There are many IoT applications and it continues to increase.Some of these IoT applications currently penetrating the technological worldinclude the following.

2.7.1 Smart Home Applications

Smart homes are probably the most common and have revolved around the Internet for a long time. Smart home devices gather and disseminate information with one another in an integrated platform and automate their response actions based on the owner's preference. Building smart home automation using IoT helps us manage our lives but as the technologies advances, the main aim of the manufacturers and designers changes their working principles into reducing the controlling and monitoring methods for all functions of the smart applications. The idea cloud IoT comes into our lives as a boon and enables all devices to use Internet connection and cloud storing platforms to work wirelessly and sometimes operation-based. When sensors are attached to them, as part of physical layer the appliances designed by the intent of the smart home concept are able to be used without touching. The systems are controlled with the help of the applications on smartphones or other devices as well as with the voice recognition option.

2.7.2 Smart Health Care

There is a huge applicability of technology, data, and communication methodologies to improve health-care solutions through telehealth and telemedicine. Health-care queries is emerging across the globe, there is a growing number of patients being remotely treated globally. The smart IoT wearables like fitness bands and blood pressure monitors to help patient's health. Numerous alert mechanisms put in smart devices to notify doctors or family members in the case of emergencies. For the physicians, IoT revolutionized their dealing with patient as it becomes easy to get into the history of a patient and access real-time health data easily. Efficiency of clinical trials increased day-by-day with real-time health data. Different health insurance companies, nowadays, collect data through IoT devices and store them in the cloud, to track the routine activities of a patient, whether they are obeying to their treatment plans or even looking into the operation processes. These smart systems automate the workflow and provide effective health-care services to the patients.

2.7.3 Smart Cities

Smart city is a futuristic powerful application of IoT creating huge curiosity among the world's population. Smart scrutiny, computerized transportation, smarter energy management systems, smart water distribution, urban security, and environmental monitoring all are examples of IoT applications for smart cities. In these cities, IoT with collaboration of other enabling technologies is used for several reasons like traffic management, public transportation, parking, utility billing, etc. With the integration of sensors, GPS data collection with cloud platforms, it will be stress-free to monitor traffic conditions of a specific area, plan construction program by predicting its influence on traffic, and finding new alternative routes whenever necessary.

2.7.4 Smart Waste Management

One of the important IoT application is selecting the right route for garbage trucks. With powerful smart waste management, IoT applications can notify truck drivers about filled dustbins and set a route for them so that they do not have to waste time by exploring locations with empty dustbins. These devices and integrated platforms also help in developing smart bins, that is, trash bins that can segregate waste into categories like plastic, metal, glass, or paper.

2.7.5 Tackling Industrial Issues

In the manufacturing Industries, IoT can be used in asset management and inventory management. Implanting IoT in the manufacturing sector can help in tracking the efficiency of the systems being used, detect any errors in the machinery, detect causes of lack of efficiency, etc. IoT in the industry can help in tackling unplanned downtime and system failures can result in life-threatening situations too.

2.7.6 Outdoor Surveillance

In smart cities IoT closed-circuit TV cameras combined with AI and machine vision, automate the surveillance of streets and live streaming. These devices also provide the great line of defense mechanism for homes in indoor or outdoor environments.

2.7.7 Smart Workplace

Smart tech solutions for wearable IoT devices take us into a paperless work environment revolutionized by the COVID-19 pandemic. The augmented reality, virtual reality, and extended reality being adopted bynumber of devices to develop smart workplace for the future of the world.

2.7.8 Smart Metering and Smart Grid

Smart metering and smart grid systems change the way of energy distribution within cities. This application of IoT help consumers for smart decisions about energy consumption, transparency of electricity bills, and value added services of utility companies, A smart grid basically promises to extract information on the behaviors of consumers and electricity suppliers in an automated fashion to improve the efficiency, economics, and reliability of electricity distribution. The smart grid also paves new ways to allow real-time data monitoring and electricity demand. These applications involve computer intelligence for the efficient resources management, outage management, load distribution, and fault detection and repairs.

2.7.9 Smart Farming

Smart farming has gain popularity and potential; that farmers can used for applications for optimizing a lot of diverse activities. These applications increase the efficiency of precise farming operations and reduced the labor extent of farming process. Smart farms can revolutionize the agriculture industry and to boost automated tasks and crop quality and quantity.

2.7.10 Smart Tracking and Monitoring System

Asset tracking is an important part of some businesses Global Positioning System (GPS) or radio frequency (RF) are used to track and monitor assets and equipment. The smart devices can be used for long-range identification and verification of assets; which is highly desirable as the online market is growing fast.

The applications of IoT with cloud technologies are numerous; and touched all aspects of life; because it is adjustable to almost any technology that is capable of providing relevant information about its own operation, about the performance of an activity and even about the environmental conditions that we need to monitor and control at a distance. Some of their key enabling services are listed in Table 2.3.

Table 2.3 Key service areas of smart cloud IoT.

S.no.	Application Area	Key Services
1	Smart Home Applications	Smart thermostats, smart lighting, smart locks, solar power surfaces, etc.
2	Smart Health Care	Provide medical coverage of patients by remote monitoring and preventive care
3	Smart Cities	Provides the smart cities ideas and operations using real-time applications
4	Smart Waste Management	Optimize driver routes and schedules to pick up community bins and replace with empty ones
5	Tackling Industrial Issues	Evolve factory ecosystem with IoT, security assessment with IoT, implant manufacturing automation
6	Outdoor Surveillance	Public safety, connected cars, traffic management
7	Smart Workplace	Multi-location office management, employee on-boarding, restricted access and control
8	Smart Metering and Smart Grid	Provide interface between you and your energy provider, enables energy management System (EMS), helping to balance the energy load in your area
9	Smart Farming	Implement smart irrigation, autonomous harvesting, smart pest management
10	Smart Tracking and Monitoring System	Pinpoint the location of any entity, device management, trip history, schedule preventative maintenance, and reminders

2.8 Conclusion

The future of smart world largely impacted by cloud computing and IoT integration. Since, the adoption of the cloud IoT paradigm enabled several new applications, many of issues are also arising. Unfortunately, there's no one-size-fits-all answer when it comes to the cloud integration with IoT. Since in some cases, monolithic collaborative applications are the best fit; while micro services offer a ton of promise to agile implementation and development, and the advent of real-time serverless integration also revolutionized and changed the working principles of huge heterogeneous IoT enabled applications by going to create an opportunity to redefine the complete application stack, and the way software is written and applications are built. As these tech bond integration trends; those who are still struggling with how to adapt to endless recent innovations such as cloud, containers and micro services, serverless computing may appear to be yet another headache to endure and be a part of an existential mistake. The critical issues and their appropriate solutions identify recent research directions in the field of cloud IoT. The collaborated cloud-IoT application scenarios also poses important research challenges such as the devices and technologies has heterogeneous in nature; their performance, reliability, scalability, security, and privacy preservation. The advantages of transferring IoT services to the cloud may depend on the necessities and boundaries of the specific use case. In spite of many concerns when standards and regulations were accepted worldwide; cloud IoT revolutionized the smart world.

Bibliography

1 Atzori, L., Iera, A., and Morabito, G. (2010). The Internet of Things: a survey. *Computer Networks* 54 (15): 2787–2805.

2 Sfar, A.R., Zied, C., and Challal, Y. (2017). A systematic and cognitive vision for IoT security: a case study of military live simulation and security challenges. *Proc. 2017 International Conference on Smart, Monitored and Controlled Cities (SM2C)*, Sfax, Tunisia, 17–19.

3 Gatsis, K. and Pappas, G.J. (2017). Wireless control for the IoT: power spectrum and security challenges. *Proc.2017 IEEE/ACM Second International Conference on Internet-of-Things Design and Implementation (IoTDI)*, Pittsburg, PA, USA, 18–21 April 2017, 22 (7), 97–114.

4 Čolaković, A. and Hadžiali, M. (2018). Internet of Things (IoT): a review of enabling technologies, challenges, and open research issues. doi: 10.1016/j.comnet.2018. 07.017.

5 Cloud Computing. Wikipedia. (May 2021). Available at http://en.wikipedia.org/ wiki/Cloud_computing.

6 Foster, I., Zhao, Y., Raicu, I., and Lu, S. (2008). Cloud Computing and Grid Computing 360-Degree Compared. *Grid Computing Environments, Workshop, 2008. GCE '08*, 10, 1.

7 Kutzias, D., Falkner, J., and Kett, H. On the complexity of cloud and IoT integration: architectures, challenges and solution approaches. *Proceedings of the 4th International Conference on Internet of Things, Big Data and Security (IoTBDS 2019)*, 376–384. doi: 10.5220/0007750403760384, ISBN: 978–989–758–369.

8 Satyanarayanan, M. (2017). The emergence of edge computing. *Computer* 50 (1): 30–39.

9 Shaukat, U., Ahmed, E., Anwar, Z., and Xia, F. (2015). Cloudlet deployment in local wireless networks: motivation, architectures, applications, and open challenges. *Journal of Network and Computer Applications*, 62: 18–40. December 2015. DOI: 10.1016/j.jnca.2015.11.009.

10 Yuan, A., Peng, M., and Zhang, K. (2018). Edge computing technologies for Internet of Things: a primer, digital communications and networks. *Digital Communications and Networks*, 4 (2): 77–86.

11 https://iot-epi.eu/wp-content/uploads/2018/07/Advancing-IoT-Platform-Interoperability-2018-IoT-EPI.pdf (accessed May 2021).

12 Received May 3, 2019, accepted May 13, 2019, date of publication May 16, 2019, date of current version May 30, 2019. Doi: 10.1109/ACCESS.2019.2917387.

13 Khanna, A. (2015). An architectural design for cloud of things. *Facta Universitatis Series: Electronics and Energetics* 29 (3): 357–365.

14 Babu, S. M., Lakshmi, A. J., and Rao, B. T. (2015). A study on cloud based Internet of Things: CloudIoT. *Global Conference on Communication Technologies (GCCT)*, 2015, 60–65.

Part 2

Quantum Computing and Its Applications

3

Quantum Computing

Principles and Mathematical Models

Arish Pitchai

Entropik Tech, 600032, Tamilnadu, Chennai, India

3.1 Introduction

Moore's law, which says integrated circuit density will double every two years, is coming to an end because of the limitations in manufacturing smaller transistors [1]. End of Moore's law doesn't mean the computers are not going to become faster anymore. Recent research reveals that a new paradigm of computation called quantum computing is capable of presenting efficient machines. Quantum computing research is interdisciplinary and demands the knowledge of mathematics, physics, and computer science to understand the working of their algorithms and to build better algorithms. Quantum computing research is experiencing rapid growth in both theoretical and practical innovations [2]. Recent developments of hardware and quantum machine learning algorithms are showing a promising future for a quantum era. This chapter provides an overview of quantum computing principles and architectures from a mathematical perspective.

The following section presents an overview to quantum computers, explaining necessary prerequisite topics from mathematics such as vector algebra and the matrix operations. Different representations of simple and composite set of qubits, the building blocks of quantum computers, are presented using Paul Dirac's bra-ket, column vector and also Bloch sphere notations. The third section contains the operations that can be applied on the two-dimensional quantum vectors. Understanding this is very important to build quantum gates. Physically measurable quantities of a sub-atomic particle are theoretically mentioned there with the help of Hermitian operators and their transformations are shown using unitary operators. Readers of the fourth chapter are shown how density operators represent both pure and mixed state quantum systems. The most unusual behavior of a quantum system can be found during the measurement phase of an algorithm. Clear explanation of measurement theory is given at the end of that section.

Any computation inside a classical computer architecture involves application of logical gates. Quantum gates, well known for their reversible property, are applied on the qubits in order to bring desirable changes to the system. Sequence of gates, called a quantum circuit, are applied on a quantum system to modify their probability amplitudes as expected by a quantum programmer. Section 5 explores the

Emerging Computing Paradigms: Principles, Advances and Applications, First Edition.
Edited by Umang Singh, San Murugesan and Ashish Seth.
© 2022 John Wiley & Sons Ltd. Published 2022 by John Wiley & Sons Ltd.

representation and construction of quantum circuits. Different combinations of single and multi-qubit gates applied on a set of qubits to perform a certain computation is termed as a quantum algorithm. Simplest quantum algorithm, named as Deutsch Algorithm, finds if a function is balance or constant faster than any other known algorithms. Deutsch Algorithms, like a "Hello World" program in C language, is explained step-by-step in Section 6. This section also covers the most admired phenomenon of quantum computing, the "Entanglement." This chapter ends with a conclusion section emphasizing the need of quantum literacy in the research community.

3.2 Quantum States and Quantum Bits

3.2.1 Vector Spaces

Bits that store either 0 or 1 at a time are the basic unit of information processing in classical computers. The quantum counterpart of bits are named as qubits (quantum bits). Classical bits are stored in a chip called RAM (Random Access Memory) before undergoing manipulations. RAM chips are made up of electronic components like transistors and capacitors. Qubits store the binary values for computation using multiple mechanisms. Photons, electrons, and semiconductor particles are some of the physical implementations of qubits. Qubits behave counter-intuitive in nature. They have the capacity to simultaneously store both one and two is a single storage location. Some pair of qubits are correlated to each other even when they are physically separated.

Vector space is the mathematical stage being used to illustrate this strange behavior of qubits. A vector space V is a non-empty set of vectors v_1, v_2, ..., v_n for which the following operations are defined:

Vector addition: $v_c = v_a + v_b$. If v_a and v_b belongs to vector space V, then the resultant vector v_c belongs to the same space V.

Scalar multiplication: $v_b = \propto v_a$. If v_a belongs to vector space V, and $\propto$ is a scalar, then v_b belongs to the same space V.

3.2.2 Qubits and Its Representations

Qubits are capable of holding both 0 and 1 simultaneously, which are labeled as $|0\rangle$ and $|1\rangle$. The $|\rangle$ notation is named as a state vector or ket vector. This state of storing multiple values at the same time is called as superposition state. It is denoted as:

$$|\psi\rangle = \alpha|0\rangle + \beta|1\rangle \tag{3.1}$$

Here α, β are complex numbers. Whenever we make a measurement of a qubit in superposition, it collapses to either $|0\rangle$ or $|1\rangle$. $|\alpha|^2$ and $|\beta|^2$ gives the probability of measuring $|0\rangle$ and $|1\rangle$ respectively. The fact that probabilities must sum to one puts the following constraint on the qubits:

$$|\alpha|^2 + |\beta|^2 = 1 \tag{3.2}$$

The vectors in a qubit vector space are represented using a column vector notation. Vector state given in Equation 3.3 is written in the following way:

$$|\psi\rangle = \begin{pmatrix} \alpha \\ \beta \end{pmatrix} \tag{3.3}$$

Coefficient of $|0\rangle$ is written in the first row and the coefficient of $|1\rangle$ in the second row. This two tuple vectors of complex numbers is denoted as $\mathbb{C}^2$.

3.2.3 Basis and Dimensions

Using the properties of vector addition and scalar multiplication, $|\psi\rangle$ in Equation 3.4 can be written as:

$$|\psi\rangle = \begin{pmatrix} \alpha \\ \beta \end{pmatrix} = \begin{pmatrix} \alpha \\ 0 \end{pmatrix} + \begin{pmatrix} 0 \\ \beta \end{pmatrix} = \alpha \begin{pmatrix} 1 \\ 0 \end{pmatrix} + \beta \begin{pmatrix} 0 \\ 1 \end{pmatrix} \tag{3.4}$$

This identifies the vectors $|0\rangle$ and $|1\rangle$ as $\begin{pmatrix} 1 \\ 0 \end{pmatrix}$ and $\begin{pmatrix} 0 \\ 1 \end{pmatrix}$. It can be seen that any vector in the qubit vector space, $\mathbb{C}^2$ can be written as a linear combination of $|0\rangle$ and $|1\rangle$. Since, one of the two vectors cannot be written as a linear combination of other, they are declared as linearly independent. Therefore, the linearly independent $|0\rangle$ and $|1\rangle$ that spans $\mathbb{C}^2$ are called as a basis. A vector space can have many basis sets. $|0\rangle$ and $|1\rangle$ is named as the computational basis of a qubit. The number of vectors in a basis set is the dimension of the vector space. Here, $\mathbb{C}^2$ is two dimensional because the basis set contains two elements $|0\rangle$ and $|1\rangle$.

3.2.4 Bloch Sphere Representation

This is a visual representation that plots the state of a qubit on the surface of a unit sphere. Following form of equation is used to denote the qubit state in a bloch sphere:

$$|\psi\rangle = \cos\frac{\theta}{2}|0\rangle + e^{i\varnothing}\sin\frac{\theta}{2}|1\rangle \tag{3.5}$$

where θ, the polar angle and $\varnothing$, the azimuthal angle are real numbers. The vector state given in Equation 3.5 can be seen in Figure 3.1.

3.2.5 Inner Product

Inner product is a mechanism used to find the length of a vector. It is very similar to the vector dot product. Two vectors of the vector space are map to one complex number. Inner product of two vectors $|u\rangle$ and $|v\rangle$ is written in Dirac's bra-ket notation as $\langle u | v \rangle$. $\langle u |$ is called the bra

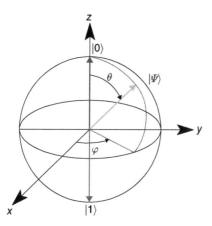

Figure 3.1 Bloch sphere.

vector and $|v\rangle$ is the ket vector. Hermitian conjugate of $|u\rangle$ is its dual $\langle u|$. It is calculated as $\langle u| = (|u\rangle)$. Square root of the inner product with itself is the norm (or length) of a vector:

$$\| u \| = \sqrt{\langle u | u \rangle} \tag{3.6}$$

If two vectors $|u\rangle$ and $|v\rangle$ are orthogonal to each other, then their inner product $\langle u|v\rangle = 0$. The vector space equipped with inner products, lengths, and angles of vectors is called Hilbert space.

3.2.6 Composite Systems and Tensor Product

A quantum computer cannot work with a single qubit in isolation. Multi-particle system is necessary to do useful computations. To understand such systems, it is necessary to be able to construct a Hilbert space H that is a composite of the independent Hilbert spaces H_1 and H_2 that are associated with each individual particle. It is written using the operator $\otimes$ as:

$$H = H_1 \otimes H_2 \tag{3.7}$$

The major tool required to perform this task is named the Kronecker or tensor product. Let $|u\rangle \in H_1$ and $|v\rangle \in H_2$ be two vectors, then the composite state $|\psi\rangle \in H_1 \otimes H_2$ is written in the following way:

$$|\psi\rangle = |u\rangle \otimes |v\rangle \tag{3.8}$$

The dimension of this composite state is the product of dimensions of Hilbert spaces H_1 and H_2. Let,

$$|u\rangle = \begin{pmatrix} \alpha_1 \\ \beta_1 \end{pmatrix} \text{ and } |v\rangle = \begin{pmatrix} \alpha_2 \\ \beta_2 \end{pmatrix}$$

Then, computing the column vector of a tensor product is given by

$$|\psi\rangle = |u\rangle \otimes |v\rangle = \begin{pmatrix} \alpha_1 \\ \beta_1 \end{pmatrix} \otimes \begin{pmatrix} \alpha_2 \\ \beta_2 \end{pmatrix} = \begin{pmatrix} \alpha_1 \alpha_2 \\ \alpha_1 \beta_2 \\ \beta_1 \alpha_2 \\ \beta_1 \beta_2 \end{pmatrix} \tag{3.9}$$

Computation in a quantum computer is possible only if the values stored in qubits are changed. A set of useful matrices that does this task are called the operators.

3.3 Quantum Operators

3.3.1 Observable

A quantum operator $\hat{A}$ is the mechanism that transforms a ket vector $|u\rangle$ into another ket vector $|v\rangle$:

$$\hat{A}|u\rangle = |v\rangle \tag{3.10}$$

Operators can also transform bra vectors:

$$\langle u | \hat{A} = \langle v |$$ (3.11)

Identity operator is a well-known operator:

$$\hat{I} | u \rangle = | u \rangle$$ (3.12)

Zero operator converts all states into zero vector:

$$\hat{0} | u \rangle = | u \rangle$$ (3.13)

Dynamic properties of a system like position, momentum, energy, etc. are called observables in quantum theory. One of the quantum postulates states that there is an operator that corresponds to the measurement of each physical observable. The product of a ket vector $| u \rangle$ and a bra vector $\langle v |$ is called an outer product and it is written as $| u \rangle \langle v |$. This outer product is a representation of an operator.

3.3.2 Pauli Operators

A set of operators that are of most importance in quantum computing are Pauli operators. $\hat{I}$, $\hat{X}$, $\hat{Y}$ and $\hat{Z}$ constitutes Pauli operators, that are also denoted by σ_0, σ_1, σ_2 and σ_3 or by σ_0, σ_x, σ_y and σ_z. Identity operator $\hat{I}$ doesn't make any change when acted on the qubit state, as given below:

$$\sigma_0 | 0 \rangle = | 0 \rangle, \sigma_0 | 1 \rangle = | 1 \rangle$$ (3.14)

Outer product version of identity operator is $\hat{I} = | 0 \rangle \langle 0 | + | 1 \rangle \langle 1 |$. Action of this outer product identity operator is given below:

$$\hat{I} | 0 \rangle = \left(| 0 \rangle \langle 0 | + | 1 \rangle \langle 1 | \right) | 0 \rangle$$
$$= | 0 \rangle \langle 0 | 0 \rangle + | 1 \rangle \langle 1 | 0 \rangle$$

It is already shown that inner product of orthogonal vectors is 0 and inner product of same vectors is related to it's length. Since computational vectors are orthogonal to each other and they are of unit length,

$$\langle 0 | 0 \rangle = \langle 1 | 1 \rangle = 1 \text{ and } \langle 0 | 1 \rangle = \langle 1 | 0 \rangle = 0$$ (3.15)

Therefore,

$$\hat{I} | 0 \rangle = | 0 \rangle.1 + | 1 \rangle.0 = | 0 \rangle$$ (3.16)

Operator $\hat{X}$, also denoted by σ_x or σ_1 has the nature of flipping the computational basis.

$$\sigma_1 | 0 \rangle = | 1 \rangle, \sigma_1 | 1 \rangle = | 0 \rangle$$ (3.17)

This is the reason researchers sometimes call $\hat{X}$ operator as quantum NOT operator.

Application of $\hat{Y}=\sigma_2=\sigma_y$ and $\hat{Z}=\sigma_3=\sigma_z$ on computational basis is as follows:

$$\sigma_2|0\rangle=-i|1\rangle, \sigma_2|1\rangle=i|0\rangle \tag{3.18}$$

$$\sigma_3|0\rangle=|0\rangle, \sigma_3|1\rangle=-|1\rangle \tag{3.19}$$

Outer product versions of $\hat{X}$, $\hat{Y}$ and $\hat{Z}$ are written below:

$$\hat{X}=|1\rangle\langle0|+|0\rangle\langle1| \tag{3.20}$$

$$\hat{Y}=-i|1\rangle\langle0|+i|0\rangle\langle1| \tag{3.21}$$

$$\hat{Z}=|0\rangle\langle0|-|1\rangle\langle1| \tag{3.22}$$

3.3.3 Matrix Representation of Operators

State of a qubit is a ket vector $|v\rangle$ that can be represented using a column matrix as follows:

$$|v\rangle=\begin{pmatrix}\alpha\\\beta\end{pmatrix} \tag{3.23}$$

Since bra vector $\langle v|$ is the hermitian conjugate of the column matrix $|v\rangle$, it can be written as a row vector as given below:

$$\langle v|=\begin{pmatrix}\alpha^* & \beta^*\end{pmatrix} \tag{3.24}$$

This shows that an outer product can be easily translated into matrix form. Consider two qubits $|u\rangle$ and $|v\rangle$.

$$|u\rangle=\begin{pmatrix}\alpha_1\\\beta_1\end{pmatrix} \text{ and } |v\rangle=\begin{pmatrix}\alpha_2\\\beta_2\end{pmatrix} \tag{3.25}$$

Matrix form of the operator $|u\rangle\langle v|$ is arrived using:

$$|u\rangle\langle v|=\begin{pmatrix}\alpha_1\\\beta_1\end{pmatrix}\begin{pmatrix}\alpha_2^* & \beta_2^*\end{pmatrix}=\begin{pmatrix}\alpha_1\alpha_2^* & \alpha_1\beta_2^*\\\beta_1\alpha_2^* & \beta_1\beta_2^*\end{pmatrix} \tag{3.26}$$

It is evident from Equation 3.26 that the operators that act on a single qubit are of size 2×2.

Matrix representation of the Pauli matrices X,Y and Z with respect to computational basis are:

$$X=\begin{pmatrix}0 & 1\\1 & 0\end{pmatrix}, Y=\begin{pmatrix}0 & -i\\i & 0\end{pmatrix} \text{ and } Z=\begin{pmatrix}1 & 0\\0 & -1\end{pmatrix} \tag{3.27}$$

Another matrix representation of quantum operators does exist with respect to computational basis. Following arrangement of elements is mandatory to construct such matrix for an arbitrary operator A.

$$A = \begin{pmatrix} \langle 0|A|0 \rangle & \langle 0|A|1 \rangle \\ \langle 1|A|0 \rangle & \langle 1|A|1 \rangle \end{pmatrix} \tag{3.28}$$

Action of Pauli Z is restated here to understand the construction of its matrix form. $Z|0\rangle = |0\rangle$ and $Z|1\rangle = -|1\rangle$. Z can be written using Equation 3.28,

$$Z = \begin{pmatrix} \langle 0|\ Z|0 \rangle & \langle 0|Z|1 \rangle \\ \langle 1|\ Z|0 \rangle & \langle 1|\ Z|1 \rangle \end{pmatrix}$$

$$= \begin{pmatrix} \langle 0|(Z|0\rangle) & \langle 0|(Z|1\rangle) \\ \langle 1|(Z|0\rangle) & \langle 1|(Z|1\rangle) \end{pmatrix}$$

$$= \begin{pmatrix} \langle 0|(|0\rangle) & \langle 0|(-|1\rangle) \\ \langle 1|(|0\rangle) & \langle 1|(-|1\rangle) \end{pmatrix}$$

$$= \begin{pmatrix} \langle 0|0\rangle & -\langle 0|1\rangle \\ \langle 1|0\rangle & -\langle 1|1\rangle \end{pmatrix}$$

$$= \begin{pmatrix} 1 & 0 \\ 0 & -1 \end{pmatrix}$$

3.3.4 More Quantum Operators

Two major classes of operators that are present in almost all quantum algorithms are Hermitian operators and Unitary operators. Operators that are their own Hermitian conjugates are the Hermitian operators.

$$H^\dagger = H \tag{3.29}$$

In quantum mechanics, all operators that represents measurable variables are Hermitian operators. An operator is called as Unitary if its adjoint is equal to its inverse: $U^\dagger = U^{-1}$. Unitary operators describe the change in state of the system with respect to time. Pauli operators are both Unitary and Hermitian.

3.3.5 Quantum Mechanics Postulates

Every system has a set of rules to govern their working principles. Framework of quantum computing paradigm too has a such rules, called as the postulates of quantum mechanics. These are a set of axioms that define how the theory behind computation works. An informal description of some of the important postulates are given here:

Postulate 1: The state of a quantum system is a time-dependent vector $|\psi\rangle(t)$ in a Hilbert space specified by a wave function. It contains all information about the system.

Postulate 2: To every physically measurable dynamical variable A in classical mechanics, there corresponds a Hermitian operator A.

Postulate 3: When the state of a qubit is measured, outcome of the measurement of a dynamical variable A is one of the eigenvalues an of the operator A corresponding to that variable.

Postulate 4: The time evolution of a closed (i.e., isolated from external environment) quantum system is governed by the famous Schrödinger equation:

$$i\hbar \frac{\partial}{\partial x}|\psi\rangle = H|\psi\rangle \tag{3.30}$$

Where H is the Hamiltonian of the system that corresponds to the total energy of the system. State of the system at time t is derived from:

$$|\psi(t)\rangle = e^{-iHt/\hbar}|\psi(0)\rangle \tag{3.31}$$

$U = e^{-iHt/\hbar}$ is the unitary operator that governs the evolution of a quantum system.

3.4 Mixed States and Measurements

3.4.1 Mixed Quantum States

When a quantum system is in definite state like $|\psi\rangle = \alpha|0\rangle + \beta||1\rangle$, it is said to be in pure state. In most of the situations, a large number of quantum systems are employed to solve complex problems. This collection of systems is called as an ensemble or mixed state system. Elements of an ensemble is found in one of the various different quantum states. There is a probability associated with each of these states. For better understanding, consider two-dimensional basis vectors $|u\rangle$ and $|v\rangle$. Prepare a collection of N systems where each system can be in of the two states $|s_1\rangle$ and $|s_2\rangle$ which are described in Equation 3.32.

$$|s_1\rangle = \alpha_1|u\rangle + \beta_1|v\rangle \text{ and } |s_2\rangle = \alpha_2|u\rangle + \beta_2|v\rangle \tag{3.32}$$

Usual rules of quantum mechanics apply on the two states. Out of N systems, $n1$ are prepared to be in state $|s_1\rangle$ and $n2$ are in state $|s_2\rangle$. Therefore, $N = n_1 + n_2$. If an element is randomly picked from this ensemble, then probability of finding that element in state $|s_1\rangle$ and $|s_2\rangle$ are n_1/N and n_2/N, respectively. The above example clearly shows the difference between composite systems and mixed state systems. Tensor product is used to represent composite systems and density operators are used to represent ensembles.

3.4.2 Density Operator

Density operator, denoted by ρ, aids in describing the statistical mixture of quantum systems. It can also be used to describe pure states. Usually, density operator of a pure state $|\psi\rangle$ is written as:

$$\rho = |\psi\rangle\langle\psi| \tag{3.33}$$

To represent an ensemble using density operator, a density operator for each of it's individual state should be constructed. Then, the probability of each such individual state is found and the possibilities should be summed up. Again, consider the two-dimensional basis vectors $|u\rangle$ and $|v\rangle$ for this example. Prepare an ensemble of N systems where each system can be in one of the two states $|s_1\rangle$ and $|s_2\rangle$, which are described in Equation 3.34.

$$|s_1\rangle = \alpha_1 |u\rangle + \beta_1 |v\rangle \text{ and } |s_2\rangle = \alpha_2 |u\rangle + \beta_2 |v\rangle \tag{3.34}$$

As mentioned in the previous section, out of N systems, n_1 are prepared to be in state $|s_1\rangle$ and $n2$ are in state $|s_2\rangle$. The density operator of each individual states are:

$$\rho_{s_1} = |s_1\rangle\langle s_1| \text{ and } \rho_{s_2} = |s_2\rangle\langle s_2| \tag{3.35}$$

Since the probabilities of finding that element in state $|s_1\rangle$ and $|s_2\rangle$ are n_1/N and n_2/N respectively, the density operator of the mixed state ensemble is:

$$\rho = (n_1/N)|s_1\rangle\langle s_1| + (n_2/N)|s_2\rangle\langle s_2| \tag{3.36}$$

Quantum researchers claim that mixed states are incoherent and pure states exhibit coherence. Quantum coherence is the property of showing relationship between qubits. This results in density operators with zero as off-diagonal elements for mixed state and non-zero off-diagonal elements for pure states most of the time. There exists another mathematical verification to characterize a mixed state. Trace (Tr) of a matrix is the sum of its diagonal elements. Trace of the square of density operator is always less than one for mixed state and equal to one for pure state systems. In summary:

- $\rho_{ij} = 0$ when $i \neq j$ and $Tr(\rho^2) < 1$ for mixed state density operators.
- $\rho_{ij} \neq 0$ when $i \neq j$ and $Tr(\rho^2) = 1$ for pure state density operators.

3.4.3 Measurement Theory

State of a quantum system is measured to find the values of it's physical properties. In quantum mechanics, state of the system undergoes a change after a measurement is made. This doesn't happen in classical systems. When a measurement is made on the state $|\psi\rangle = \alpha |0\rangle + \beta |1\rangle$, it is forced to collapse into $|0\rangle$ or $|1\rangle$ after measurement, i.e., the original state of the system is lost. This is because, the qubit being measured happens to contact the measuring device, a larger environment called ancilla. Quantum systems that are isolated from larger environment are called closed systems and others are open systems.

Projective or von Neumann measurement is one of the oldest quantum measurement models. Mutually exclusive properties of a qubit like energy state, position, etc. are measured using projective measurements. A projection operator P, which is hermitian, is used in the measurement process. $P = P$ and $P^2 = P$ are the mathematical properties of a projection operator. If product of any two projection operators $P1$ and $P2$ is zero, then they are orthogonal to each other. Number of orthogonal projection operators in an n-dimensional Hilbert space is less than or equal to n.

In a two-dimensional state $|\psi\rangle = \alpha |0\rangle + \beta |1\rangle$, the possible projection operators are $P_0 = |0\rangle\langle 0|$ and $P_1 = |1\rangle\langle 1|$. When a measurement is made, the probability of finding i^{th} outcome is:

$$Pr(i) = |P_i|\psi\rangle|^2 = (P_i|\psi\rangle)^\dagger (P_i|\psi\rangle) = \langle\psi|P_i^2|\psi\rangle \tag{3.37}$$

Measurement outcome of a quantum state $|\psi\rangle$ is one of the eigenvalues of the observable A with eigenvectors $|u_i\rangle$, each with eigenvalues a_i. Operator associated with the observable A can be decomposed as:

$$A = \sum_{i=1}^{n} a_i |u_i\rangle \langle u_i| = \sum_{i=1}^{n} a_i P_i \tag{3.38}$$

Now, expansion of the qubit state $|\psi\rangle$ with respect to the eigenvectors of observable A can be done as given in Equation 3.39.

$$|\psi\rangle = \sum_{i=1}^{n} (\langle u_i \| \psi\rangle) |u_i\rangle = \sum_{i=1}^{n} c_i |u_i\rangle \tag{3.39}$$

Probability $Pr(i)$ of getting the measurement result as a_i is $|c_i|^2 = |\langle u_i \| \psi\rangle|^2$. This result is derived from Born rule, which is a famous rule of quantum mechanics.

It is already mentioned that the system collapses and changes its physical state after performing a measurement on it. Mathematically, the collapsed state of the system $|\psi'\rangle$ is calculated using Equation 3.40.

$$|\psi'\rangle = \frac{P_i |\psi\rangle}{\sqrt{\langle \psi | P_i | \psi\rangle}} \tag{3.40}$$

3.4.4 Measuring Composite Systems

Most of the quantum algorithms involve composite system of qubits. To apply operators on a composite system, it is necessary to formulate a method to combine the operators together. Tensor product is used to represent multiple qubits in a single column. The same tensor product can be used to combine two or more operators. Consider a couple of operators A and B on a two-dimensional Hilbert space. Let,

$$A = \begin{pmatrix} a_{11} & a_{12} \\ a_{21} & a_{22} \end{pmatrix} \text{ and } B = \begin{pmatrix} b_{11} & b_{12} \\ b_{21} & b_{22} \end{pmatrix} \text{ be the matrix form of the two operators. Tensor}$$

product of the operators can be constructed using Equation 3.41.

$$A \otimes B = \begin{pmatrix} a_{11}B & a_{12}B \\ a_{21}B & a_{22}B \end{pmatrix} = \begin{pmatrix} a_{11}b_{11} & a_{11}b_{12} & a_{12}b_{11} & a_{12}b_{12} \\ a_{11}b_{21} & a_{11}b_{22} & a_{12}b_{21} & a_{12}b_{22} \\ a_{21}b_{11} & a_{21}b_{12} & a_{22}b_{11} & a_{22}b_{12} \\ a_{21}b_{21} & a_{21}b_{22} & a_{22}b_{21} & a_{22}b_{22} \end{pmatrix} \tag{3.41}$$

When measuring a two-qubit composite system using projection operators $P1$ on first qubit and $P2$ on the second qubit, tensor product $P_1 \otimes P_2$ is constructed. If measurement is to be made on only the first qubit, then the tensor product $P_1 \otimes I$ is applied on the composite system. The Hilbert space of composite systems not only contains product states like $|u\rangle \otimes |v\rangle$. It also contains non-separable states that will be discussed soon in this chapter.

3.5 Constructing Circuits

3.5.1 Classical Logic Gates

In a classical computer, information stored as bits is either transferred or processed. To process the information, logic gates are very much essential. Simplest of the logic gates is NOT gate that flips the input. If the input is zero, then the output is one and vice-versa. Every logic gate has a truth table associated with it. Table 3.1 shows the behavior of classical NOT gate.

NOT gate takes one input and sends out one output. There are gates like OR, AND, and XOR (exclusive-OR) which takes two inputs and produces one output. The two inputs are labeled as A and B. If either A or B is one, then the output of OR gate is a 1, and is 0 otherwise. For better insight, a truth table of OR gate is given in Table 3.2.

Another important classical gate to be noted is XOR. It produces 1 as an output when any one of the inputs is 1 and the other is 0. XOR operation is denoted using the $\oplus$ symbol. Truth table of XOR is given in Table 3.3.

There are many interesting classical gates available. Most of them are irreversible, i.e., input cannot be determined just by looking at the output produced. On the contrary, quantum gates are reversible. One of the few reversible classical gates is Fredkin gate. This gate takes three inputs, out of which first input is called as a control bit (C). Control bit determines whether the function can be applied or not. In other words, if $C = 0$, then the input bits simply pass through the gate without getting affected. If $C = 1$, then the bits sent as input gets interchanged as shown in Table 3.4.

Table 3.1 Truth table of classical NOT gate.

INPUT	OUTPUT
A	NOT(A)
0	1
1	0

Table 3.2 Truth table of classical OR gate.

INPUT		OUTPUT
A	B	A OR B
0	0	0
0	1	1
1	0	1
1	1	1

Table 3.3 Truth table of classical XOR gate.

INPUT		OUTPUT
A	B	A $\oplus$ B
0	0	0
0	1	1
1	0	1
1	1	0

Table 3.4 Truth table of Fredkin gate.

INPUT			OUTPUT	
C	A	B	A'	B'
0	0	0	0	0
0	0	1	0	1
0	1	0	1	0
0	1	1	1	1
1	0	0	0	0
1	0	1	1	0
1	1	0	0	1
1	1	1	1	1

3.5.2 Single Qubit Gates

Quantum gates when applied transforms the state of a qubit. Gates in the quantum world are unitary operators. It is previously shown that for unitary operators, their adjoints are equal to their inverses. If U is a unitary operator, then $U^{\dagger}=U^{-1}$, thus, $UU^{\dagger}=U^{\dagger}U=I$. Since, all quantum operators have a matrix form, all quantum gates can be represented using matrices. If a quantum gate is applied on a single qubit, then the dimension of the gate matrix is 2×2.

A single quantum gate, which was introduced earlier in this chapter, is the quantum NOT gate (or Pauli's X operator).

$$X = U_{NOT} = \begin{pmatrix} 0 & 1 \\ 1 & 0 \end{pmatrix} \tag{3.42}$$

When a quantum NOT is applied on the computational basis $|0\rangle$ or $|1\rangle$, the state gets flipped. That is, $|0\rangle \mapsto |1\rangle$ and $|1\rangle \mapsto |0\rangle$. It is known that $|0\rangle = \begin{pmatrix} 1 \\ 0 \end{pmatrix}$ and $|1\rangle = \begin{pmatrix} 0 \\ 1 \end{pmatrix}$. Application of U_{NOT} on the computational basis is mathematically explained in Equations 3.43 and 3.44 here:

$$U_{NOT} |0\rangle = \begin{pmatrix} 0 & 1 \\ 1 & 0 \end{pmatrix} \begin{pmatrix} 1 \\ 0 \end{pmatrix} = \begin{pmatrix} 0 \\ 1 \end{pmatrix} = |1\rangle \tag{3.43}$$

$$U_{NOT} |1\rangle = \begin{pmatrix} 0 & 1 \\ 1 & 0 \end{pmatrix} \begin{pmatrix} 0 \\ 1 \end{pmatrix} = \begin{pmatrix} 1 \\ 0 \end{pmatrix} = |0\rangle \tag{3.44}$$

The way U_{NOT} acts on an arbitrary state $|a\rangle$ can also be written as $U_{NOT} |a\rangle = |a \oplus 1\rangle$, where $\oplus$ is the classical XOR operator. In the previous section, XOR gate was explained with its truth table. It produces 1 as an output if and only if one of the inputs is 0 and the other is 1. So, if $|a\rangle = 0$, then $U_{NOT} |0\rangle = |0 \oplus 1\rangle = |1\rangle$ and if $|a\rangle = 1$, then $U_{NOT} |1\rangle = |1 \oplus 1\rangle = |0\rangle$.

As it is known that all the Pauli operators are unitary, they come under the single qubit gates category. Pauli Z is called the phase flip gate because it inverts the phase of the quantum state. Hadamard gate H plays a vital role in many quantum algorithms. Matrix form of H gate is $\frac{1}{\sqrt{2}} \begin{pmatrix} 1 & 1 \\ 1 & -1 \end{pmatrix}$. S, T and R($\phi$) gates are other frequently used single qubit gates.

3.5.3 Basics of Quantum Circuit

Circuit representation of application of quantum gates helps in programming quantum computers with ease. It is very similar to that of classical circuits. A classical NOT gate application is shown in Figure 3.2.

Single qubit quantum gates like I, Pauli X, Y, and Z operators are pictorially represented using a block with an alphabet representing the gate in it. An input line is drawn to the left of the block and an output line to its right as shown in Figure 3.3.

Another important gate that completes a circuit is the measurement gate. Circuit representation of measurement gate is given in Figure 3.4.

There are variety of quantum gates with different pictorial representations available. Each of them will be discussed with appropriate circuit diagrams in future sections of this chapter.

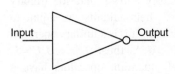

Input ——— Output

Figure 3.2 Circuit diagram of classical NOT gate.

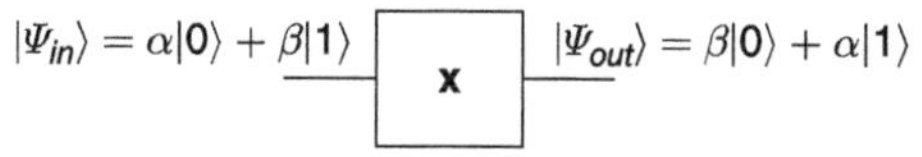

$|\Psi_{in}\rangle = \alpha|0\rangle + \beta|1\rangle$ 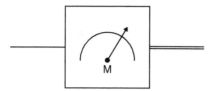 $|\Psi_{out}\rangle = \beta|0\rangle + \alpha|1\rangle$

Figure 3.3 Circuit diagram of quantum NOT gate.

Figure 3.4 Circuit diagram of quantum measurement gate.

3.5.4 Controlled Quantum Gates

Controlled gates helps in implementing conditional programming structures like *if-else* construct. It works similar to that of a classical Fredkin gate, which is explained earlier in this chapter. The name comes from the inclusion of a control bit C. A controlled gate does nothing on the qubit if control bit C is equal to 0 and it does a specified operation if the control bit C is set to 1. The first control gate to be explored here is the CNOT (Controlled NOT) gate. Circuit diagram of CNOT is shown in Figure 3.5.

First qubit input of the CNOT gate acts as the control qubit, which is not altered during the application of the gate. The second qubit is called the target qubit, which gets flipped when the control qubit is 1. The NOT operation is written using the XOR representation. $|0\rangle|0\rangle$, $|0\rangle|1\rangle$, $|1\rangle|0\rangle$ and $|1\rangle|1\rangle$ are the possible states a two-qubit composite system can possess. Action of a CNOT gate on the possible states are given in Table 3.5.

The matrix representation of CNOT gate with respect to the states $|00\rangle$, $|01\rangle$, $|10\rangle$, and $|11\rangle$ basis is given in Equation 3.45.

$$\text{CNOT} = \begin{pmatrix} 1 & 0 & 0 & 0 \\ 0 & 1 & 0 & 0 \\ 0 & 0 & 0 & 1 \\ 0 & 0 & 1 & 0 \end{pmatrix} \tag{3.45}$$

Column vector forms of $|00\rangle$, $|01\rangle$, $|10\rangle$ and $|11\rangle$ are $\begin{pmatrix} 1 \\ 0 \\ 0 \\ 0 \end{pmatrix}$, $\begin{pmatrix} 0 \\ 1 \\ 0 \\ 0 \end{pmatrix}$, $\begin{pmatrix} 0 \\ 0 \\ 1 \\ 0 \end{pmatrix}$ and $\begin{pmatrix} 0 \\ 0 \\ 0 \\ 1 \end{pmatrix}$, respectively. Action of matrix form of CNOT on the 2-qubit column vectors are given below:

$$\text{CNOT}|00\rangle = \begin{pmatrix} 1 & 0 & 0 & 0 \\ 0 & 1 & 0 & 0 \\ 0 & 0 & 0 & 1 \\ 0 & 0 & 1 & 0 \end{pmatrix}\begin{pmatrix} 1 \\ 0 \\ 0 \\ 0 \end{pmatrix} = \begin{pmatrix} 1 \\ 0 \\ 0 \\ 0 \end{pmatrix} = |00\rangle,$$

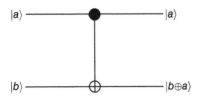

Figure 3.5 Circuit diagram of CNOT gate.

Table 3.5 Action of CNOT on a 2-qubit composite system.

INPUT		OUTPUT	
Control qubit $|a\rangle$	Target qubit $|b\rangle$	Control qubit $|a\rangle$	Target qubit $|b \oplus a\rangle$
$|0\rangle$	$|0\rangle$	$|0\rangle$	$|0\rangle$
$|0\rangle$	$|1\rangle$	$|0\rangle$	$|1\rangle$
$|1\rangle$	$|0\rangle$	$|1\rangle$	$|1\rangle$
$|1\rangle$	$|1\rangle$	$|1\rangle$	$|1\rangle$

$$\text{CNOT}|01\rangle = \begin{pmatrix} 1 & 0 & 0 & 0 \\ 0 & 1 & 0 & 0 \\ 0 & 0 & 0 & 1 \\ 0 & 0 & 1 & 0 \end{pmatrix} \begin{pmatrix} 0 \\ 1 \\ 0 \\ 0 \end{pmatrix} = \begin{pmatrix} 0 \\ 1 \\ 0 \\ 0 \end{pmatrix} = |01\rangle,$$

$$\text{CNOT}|10\rangle = \begin{pmatrix} 1 & 0 & 0 & 0 \\ 0 & 1 & 0 & 0 \\ 0 & 0 & 0 & 1 \\ 0 & 0 & 1 & 0 \end{pmatrix} \begin{pmatrix} 0 \\ 0 \\ 1 \\ 0 \end{pmatrix} = \begin{pmatrix} 0 \\ 0 \\ 0 \\ 1 \end{pmatrix} = |11\rangle, \text{ and}$$

$$\text{CNOT}|11\rangle = \begin{pmatrix} 1 & 0 & 0 & 0 \\ 0 & 1 & 0 & 0 \\ 0 & 0 & 0 & 1 \\ 0 & 0 & 1 & 0 \end{pmatrix} \begin{pmatrix} 0 \\ 0 \\ 0 \\ 1 \end{pmatrix} = \begin{pmatrix} 0 \\ 0 \\ 1 \\ 0 \end{pmatrix} = |10\rangle$$

This confirms the action of a Controlled NOT gate with clear evidence. This nature of flipping the target qubit enables CNOT gate to transform a state $\alpha|0\rangle + \beta|1\rangle$ to $\beta|0\rangle + \alpha|1\rangle$ when its control qubit is set to 1. In quantum mechanics, there are many controlled gates like CY, CZ, Toffoli, etc.

3.6 Supremacy of the Quantum Algorithms

3.6.1 Quantum Interference

An algorithm is informally defined as a set of well explained steps to perform a computation. Quantum algorithms are nothing but a series of gate operations on the given set of input qubits with some measurements at the end. Solutions to certain problems using quantum algorithms shows speed up in its execution time than their classical counterparts. This is mainly because of the inherent quality of quantum systems called the interference and superposition. An important gate that helps to achieve superposition is H gate (Hadamard). It takes $|0\rangle$ as input and converts it to $1/\sqrt{2}\left(|0\rangle+|1\rangle\right)$ and transforms input state $|1\rangle$ into $1/\sqrt{2}\left(|0\rangle-|1\rangle\right)$. Action of Hadamard matrix on the computational basis is given again in Equations 3.47 and 3.48.

$$H|0\rangle = \frac{1}{\sqrt{2}}\begin{pmatrix} 1 & 1 \\ 1 & -1 \end{pmatrix}\begin{pmatrix} 1 \\ 0 \end{pmatrix} = \frac{1}{\sqrt{2}}\begin{pmatrix} 1 \\ 1 \end{pmatrix} = \frac{1}{\sqrt{2}}\left(|0\rangle+|1\rangle\right) \tag{3.46}$$

$$H|1\rangle = \frac{1}{\sqrt{2}}\begin{pmatrix} 1 & 1 \\ 1 & -1 \end{pmatrix}\begin{pmatrix} 0 \\ 1 \end{pmatrix} = \frac{1}{\sqrt{2}}\begin{pmatrix} 1 \\ -1 \end{pmatrix} = \frac{1}{\sqrt{2}}\left(|0\rangle-|1\rangle\right) \tag{3.47}$$

As mentioned earlier, quantum gates are reversible. It means that when a Hadamard is applied to a qubit in uniform superposition $1/\sqrt{2}\left(|0\rangle+|1\rangle\right)$, the result is $|0\rangle$. In other words, when a Hadamard is applied twice, the input state will be restored. Application of H gate on an arbitrary quantum state $|\psi\rangle = \alpha|0\rangle + \beta|1\rangle$ is an example of the phenomenon quantum interference. Equation 3.48 shows the result of Hadamard on $|\psi\rangle$.

$$H|\psi\rangle = \frac{1}{\sqrt{2}}\begin{pmatrix} 1 & 1 \\ 1 & -1 \end{pmatrix}\begin{pmatrix} \alpha \\ \beta \end{pmatrix} = \frac{1}{\sqrt{2}}\begin{pmatrix} \alpha+\beta \\ \alpha-\beta \end{pmatrix} = \left(\frac{\alpha+\beta}{\sqrt{2}}\right)|0\rangle + \left(\frac{\alpha-\beta}{\sqrt{2}}\right)|1\rangle \tag{3.48}$$

Interference is the term that prevails in wave mechanics. There are two types of inferences: positive and negative. In positive interference, two wave amplitudes gets added and in negative interference, the amplitudes decreases in value. In a quantum state $|\psi\rangle$, the coefficients α and β are called as the probability amplitudes of $|0\rangle$ and $|1\rangle$, respectively. Next subsection shows the way of exploiting quantum interference to achieve quantum parallelism.

3.6.2 Quantum Parallelism in Deutsch Algorithm

Quantum parallelism is the property of evaluating a function $f(x)$ with many values of x simultaneously. Imagine a function $f(x)$ that takes one bit as input and produces a single bit output. The only two possible values of x are 0 and 1. $f(x)$ can be an identity function as in Equation 3.49.

$$f(x) = \begin{cases} 0, \text{ if } x=0 \\ 1, \text{ if } x=1 \end{cases} \tag{3.50}$$

or a bit flip function as shown in Equation 3.50.

$$f(x) = \begin{cases} 1, \text{ if } x=0 \\ 0, \text{ if } x=1 \end{cases} \tag{3.51}$$

Identity and bit flip are the examples of balance functions, where half of the output results are different from the other half. There is another class of functions called constant functions where the outputs of the function is always the same irrespective of the inputs. Two of the constant functions are shown in Equation 3.51.

$$f(x)=0 \text{ and } f(x)=1 \tag{3.52}$$

Being a constant or balanced function is a global property of a function. Deutsch's algorithm takes a function as input and determines this global property of the function. Power of quantum parallelism is demonstrated using this algorithm. Classical approach uses if-else construct to determine this global property. The input function is evaluated twice and the outputs are compared to arrive at the solution. Quantum parallelism enables us to evaluate the function simultaneously in Deutsch's algorithm.

Key element of this algorithm is the unitary operation U_f, which when applied on two qubits $|a\rangle$ and $|b\rangle$ produces two outputs. First output is the undisturbed qubit $|a\rangle$ and the second output is the XOR of $|b\rangle$ and $f(|a\rangle)$. That is,

$$U_f |a,b\rangle = |a,b \oplus f(a)\rangle \tag{3.53}$$

Two qubits $|00\rangle$ are taken H gate is applied on the first qubit. It is earlier shown that application of H gate on $|0\rangle$ produces a superposition $\frac{1}{\sqrt{2}}(|0\rangle+|1\rangle)$. Action of Uf on this superposition state gives:

$$U_f \left(\frac{|0\rangle+|1\rangle}{\sqrt{2}} \right) |0\rangle = \frac{|0,0 \oplus f(0)\rangle + |1,0 \oplus f(1)\rangle}{\sqrt{2}} \tag{3.54}$$

The above equation shows the action of quantum parallelism in evaluating $f(x)$ simultaneously for multiple values simultaneously. Circuit diagram of the above implementation is given in Figure 3.6.

Problem with a quantum computer is it is not possible to read both the values simultaneously. If an attempt to read the second qubit, where the evaluation is done, the system collapses and any one of the evaluations of $f(x)$ is revealed and the other is lost.

Deutsch's algorithm shows how to exploit the power of quantum parallelism to determine global property without missing the information in measurement. The algorithm takes two qubit composite state $|\psi_0\rangle = |01\rangle$ and applies H gate on both the qubits. Application of two H gates is denoted by $H^{\otimes 2}$ or $H \otimes H$. The resultant state is:

$$|\psi_1\rangle = (H \otimes H)|01\rangle = \left(\frac{|0\rangle+|1\rangle}{\sqrt{2}} \right)\left(\frac{|0\rangle-|1\rangle}{\sqrt{2}} \right) \tag{3.55}$$

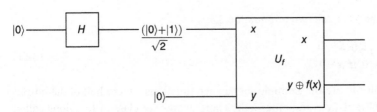

Figure 3.6 Circuit diagram to show quantum parallelism.

Next step in the algorithm is to apply U_f on the resultant state to produce:

$$|\psi_2\rangle = U_f\left(\frac{|0\rangle+|1\rangle}{\sqrt{2}}\right)\left(\frac{|0\rangle-|1\rangle}{\sqrt{2}}\right) = \begin{cases} \pm\left(\frac{|0\rangle+|1\rangle}{\sqrt{2}}\right)\left(\frac{|0\rangle-|1\rangle}{\sqrt{2}}\right) & \text{if } f(0) = f(1) \\ \pm\left(\frac{|0\rangle-|1\rangle}{\sqrt{2}}\right)\left(\frac{|0\rangle-|1\rangle}{\sqrt{2}}\right) & \text{if } f(0) \neq f(1) \end{cases}$$ (3.56)

Finally a H gate is applied on the first qubit and the second qubit is left undisturbed. The process is shown in Equation 3.57 below:

$$|\psi_3\rangle = (H \otimes I)|\psi_2\rangle = \begin{cases} \pm|0\rangle\left(\frac{|0\rangle-|1\rangle}{\sqrt{2}}\right) & \text{if } f(0) = f(1) \\ \pm|1\rangle\left(\frac{|0\rangle-|1\rangle}{\sqrt{2}}\right) & \text{if } f(0) \neq f(1) \end{cases}$$ (3.57)

Rewriting the final state $|\psi_3\rangle$ as $\pm|f(0)\oplus f(1)\rangle\left(\frac{|0\rangle-|1\rangle}{\sqrt{2}}\right)$ brings the algorithm to produce the output. If it is a constant function, then $|f(0)\oplus f(1)\rangle$ becomes $|0\rangle$. Similarly the first qubit becomes $|1\rangle$ if the function is balanced. Circuit diagram of the algorithm is shown in Figure 3.7.

3.6.3 More Quantum Algorithms

Generalization of Deutsch algorithm for multiple qubits was devised in the year 1992 by a British physicist David Deutsch and an Australian mathematician Richard Jozsa [3]. It solves the problem of finding the global property (balanced or constant) of an arbitrary function that takes multiple qubits as input. Classical solution to this problem requires exponential number of queries but a quantum computer does the job in a single query. Next notable algorithm in the field of quantum computing is the Shor's algorithm designed by the American mathematician Peter Shor in 1994 [4]. This algorithm is capable of finding the factors of a number N in polynomial time. Quantum fourier transform, the quantum counterpart of discrete fourier transform is the reason for this speed up. Invention of this algorithm created a new field of research called post quantum cryptography where researchers are busy finding a way to develop cryptographic algorithms that are quantum-resistant. Another

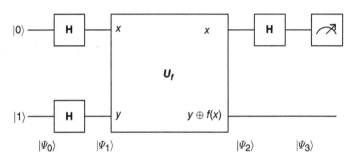

Figure 3.7 Circuit diagram of Deutsch algorithm.

famous algorithm that is capable of searching an element in an unsorted array is called the Grover's search algorithm [5]. A classical linear search has a time complexity of O(n) whereas the quantum search algorithm finds the element in $O(\sqrt{n})$ time. Researchers are busy inventing more quantum algorithms that exhibits significant speed up over their classical rivals.

3.6.4 Entanglement, Teleportation, and Superdense Coding

Entanglement is another interesting counter-intuitive property of sub-atomic particles like superposition. Consider two quantum particles X and Y entangled with each other. State of entanglement means one or more properties of system X are highly correlated with the values of the same properties of system Y. This correlation exists even when those particles are separated. This entanglement phenomenon is the major resource of quantum information technology. A couple of weird operations called teleportation and superdense coding are possible because of this inherent property of subatomic particles. Using teleportation information of a quantum state can be transmitted from person A to person B without getting tampered by adversary parties. This transmission through a communication channel happens faster than the speed of light. The superdense coding protocol can be used to communicate two bits of information using a single qubit. Individuals involved in this transmission should hold each of the two entangled particles to achieve the goal.

3.7 Conclusion

As mentioned earlier, quantum computing research is exhibiting tremendous development globally. The three main branches of research interest include quantum computing, quantum information, and quantum cryptography. Quantum computing deals with inventions of new algorithms that are capable of solving computationally complex problems by harnessing superposition phenomenon. Hybrid quantum algorithms, variational quantum solvers, and Quantum Approximation Optimization Algorithms (QAOA) are the recent breakthroughs in this field. Quantum information focuses in storage of information and its transmission through quantum channels. Free space quantum communication, the transmission of information to satellites from ground through quantum channels is one of the achievements of quantum information. Quantum cryptography studies different possible ways to secure a communication channel. Quantum computing has proved its efficiency in multi-dimensions in the past few decades through these developments.

Bibliography

1 Shalf, J. (2020). The future of computing beyond Moore's law. *Philosophical Transactions of the Royal Society A* 378 (2166): 20190061.

2 Preskill, J. (2018). Quantum computing in the NISQ era and beyond. *Quantum* 2: 79.

3 Deutsch, D. and Jozsa, R. (1992). Rapid solution of problems by quantum computation. *Proceedings of the Royal Society of London. Series A: Mathematical and Physical Sciences* 439 (1907): 553–558.

4 Shor, P. W. (November 1994). Algorithms for quantum computation: discrete logarithms and factoring. *Proceedings 35th Annual Symposium on Foundations of Computer Science*, 124–134. IEEE.

5 Grover, L. K. (July 1996). A fast quantum mechanical algorithm for database search. *Proceedings of the Twenty-Eighth Annual ACM Symposium on Theory of Computing*, 212–219.

Additional Resources

1) Neilson, M. and Chuang, I. (2010). *Quantum Computation and Quantum Information*. Cambridge University Press.

2) McMahon, D. (2007). *Quantum Computing Explained*. John Wiley & Sons.

3) https://qiskit.org/textbook/preface.html

4) https://pennylane.ai/qml/index.html

4

Quantum Cryptography and Security

Anukriti[1] and Vandana Niranjan[2]

[1]*Masters in technology in the area of VLSI and Chip design at Indira Gandhi Delhi Technical University Delhi, India*
B.Tech. degree in Electronics and Communication Engineering from Guru Gobind Singh Indraprastha University, Delhi
[2]*Professor in the Department of Electronics and Communication Engineering at Indira Gandhi Delhi Technical University Delhi, India*

4.1 Introduction to Encryption

The use of codes and ciphers to protect secrets dates back to the Roman Empire around 1900 BC [1, 2]. The concept was to portray hieroglyphic symbols instead of using usual roman characters for communicating message. The purpose of encryption (a.k.a. "Secret Writing") was to retain the secrecy of the message without letting the unwanted to understand the message, and the process has been useful ever since. The encryption is a process of encoding the information from what we call plain-text, that is the information that could be understood without any difficulty to an unusual form of information called cipher-text, that is rather too complicated to be understood in the first look.

An example of general encoding was explained by Julius Caesar [3] in 100 BC by using substitution ciphering to encode military messages (see Figure 4.1).

Modern day encryption is referred to as "cryptography," that is a process of converting ordinary plain text data into unintelligent information and vice-versa [4]. In the world of users connected over Internet, protecting information from reaching unauthentic terminals is complicated and as the number of hackers increases, it has become more important to secure our data with the most complex lock system for that only source users have a key to.

Cryptography is a developing domain with its usage in protection of various communication protocols or algorithms. An algorithm is considered secure if an attacker is unable to determine any plain text or key, given the cipher text. The confidentiality and integrity protections offered by cryptographic protocols can protect communications from malicious spying and damaging the originality of data. Authenticity protections provide assurance that users are actually communicating with the systems as intended. It can also be used to protect data at rest. Data on a removable disk or in a database can be encrypted to prevent disclosure of sensitive data should the physical media be lost or stolen. In addition, it can also provide integrity protection of data at rest to detect malicious tampering.

Being an arising technology, its integration with the laws of quantum physics upgrades the security to the quantum level, that will have a significant edge and advantage compared to classical encryption. The two parties will have secure

Emerging Computing Paradigms: Principles, Advances and Applications, First Edition.
Edited by Umang Singh, San Murugesan and Ashish Seth.

Figure 4.1 Substitution ciphering. (*Source*: Caesar Cipher). *Source*: http://www.math.stonybrook.edu/~moira/mat331-spr10/papers/1987%20LucianoCryptology%20From%20Caesar%20Ciphers%20to.pdf

communication based on the invariabilities of the laws of the quantum mechanics, making it immune to hacking.

4.1.1 Brief on Protocols

A protocol is similar to encryption, having its application in day-to-day life of intelligent beings. A protocol is a set of rules, without that a system has great possibility of failure and causing disastrous consequences. While using encryption, it is important to have a protocol for communication of information between two terminals to avoid uncertainties in the outcomes.

In modern society, most of the work is done over the Internet and a user has to send and receive hundreds of information in a day [5]. Throughout the globe, this number sums up to millions of information in a single day. Hence it is important for management aspects to have a protocol that could be useful in encryption as well as authentication of data between intended source and destination users.

A cryptographic protocol usually incorporates a key agreement or establishment whereby two or more parties can agree on a key in such a way that both influence the outcome. The key exchange systems have one user to generate the key, and simply transmit that key to the other party for decoding the message. Apart from this, the cryptographic protocols perform entity authentication, and falls under non-repudiation method category (see Figure 4.2).

There are two forms of cryptography, symmetric, and asymmetric. In the previous section we saw an example of symmetric cryptography. Public key cryptography [7] is the most abundantly used protocol in encryption, that is a form of asymmetric encryption. The concept of a public key cryptography involves two separate keys: First is a private key that is used to encrypt or decrypt the information and put a digital signature on the file. The second key is a public key that is only used to encrypt the information and verify digital signatures. Important point to note here is the private key is kept a secret and only intended users have knowledge of this key, but the public key is sent over a public channel, that implies that everyone on the channel has access to this key. So, if some hacking user is taping the network for the key, only the public key is disclosed and the hacker could not use it to decode the information as the public key does not contain information on how to decrypt the message.

Conversely, if the destination user needs to reply, then the source user's public key is used to encrypt the reply to ensure that only the intended source user can decrypt it. In this manner, the role of sender and receiver change in every conversation.

 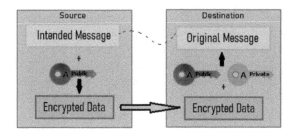

Figure 4.2 Asymmetric ciphering at source user and destination user. (*Source*: Modified from Diffie and Hellman, 1976 key exchange scheme [6]/Wikimedia Commons.)

It has now became a need to question that why symmetric encryption is still being used? The main reason behind this question is the fact that since symmetric encryption uses only single key unlike asymmetric encryption, it is faster to encrypt and decrypt information. So, if a user is sending a few page of documents or email, there is no noticeable difference. However, encrypting gigabytes or terabytes of data on a hard drive can make a big difference. Symmetric encryption is also considered stronger than asymmetric encryption, but both are sufficiently hard to break so that is not a practical issue to debate.

4.1.2 Quantum Cryptography

When the computers were in early generations, their ability to solve computational problems is considered exceptional. Most of the problems are related to arithmetic or logical queries. Classical cryptography is one such form of computational problem for classical computers. The classical cryptography is a high complexity mathematical problem for the instance factorization of large numbers [8]. Factorization is in itself a high complexity computational problem for classical computers and in cryptography it is meant to be understood as more than 10 bits of factorization. No algorithm has been designed that can factor an integer in polynomial time that can factor a n-bit number in t (time) for some constant k, such that $O\left(n^{k}\right)\epsilon\, t$ [9].

The generation of computers in that we are living today is relatively much powerful and computationally sufficient for solving most kind of problems. However, still for a problem like factorization of a 10-bit number, the most powerful supercomputer will take few hours to solve it. With the development of quantum computers it is being experimented that these computers are exponentially faster than classical ones in not all, but certain form of computation. This creates an opportunity for exceptional application for quantum machines to work on problems that are considered too complicated for classical computers.

An application of such properties is quantum cryptography [8]. To understand about the cryptographic part, the two important elements of quantum mechanics on that quantum cryptography depends are required to be understood first. They are:

a) Heisenberg uncertainty principle suggests that if a user measures one thing accurately, then it cannot measure another thing with same accuracy. The only odd thing about this principle is that it becomes true only for the instant at that the user perform a measurement of something. Photons rely on this principle due to their wave like structure. Photons are polarized or tilted in a certain direction and when a measurement is performed, all subsequent polarizations gets affected. This principle plays the vital role to prevent the efforts of attacker in quantum cryptography [10].

b) Photon polarization principle suggests that a spy user can never create a copy of the quantum bits, due to the no-cloning principle. If an unintended attempt is made for measurement, it will corrupt the information [11].

It has been understood from above sections about the cryptographic procedures and its application, one of those problems is breaking certain types of encryption, particularly the methods used in present public key infrastructure, that underlies practically all of today's online communications. Quantum computers are so inventive that with its usage, even the toughest factorization problem could be solved within minutes. Moreover, instead of solving one problem at a time, with quantum computing, one can solve multiple problems simultaneously, with the same processing power.

4.1.3 Threat to Open World Security Algorithms

The need to develop quantum cryptography is on the mind of many people and what this chapter will give insight into. The world relies on encryption to protect everything starting from financial transactions to securing databases and other sensitive information. To protect our data, computer networks, and other digital systems, the defense organizations rely on the fact that even for the world's strongest classical supercomputer, it would take more than a decade's time to decode and crack their cryptographic algorithms.

This new reality has its own benefits and drawbacks. Majority of the drawback is concerned with security of communication. As more and more systems are connecting online, more and more sensitive and confidential information is moving to the Internet, and problems like confidentiality breaches are becoming more common and significant. However, in 1977 three colleagues from Massachusetts Institute of Technology (MIT), created an algorithm used by modern computers to encrypt and decrypt messages. They named the algorithm RSA (Rivest-Shamir-Adleman) [12] and it is a form of public key cryptography (see Figure 4.3). The algorithm is based on the fact that finding the factors of a large composite number is difficult and that when the number is a prime number, the factorization becomes exponentially difficult.

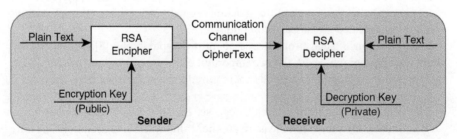

Figure 4.3 RSA communication between a sender and receiver. (*Source*: Based on Rivest et al., 1978 [12]).

Decrypting the RSA for a composite number in a small range of 10^4 becomes a hefty load for classical computer. In practical implemented RSA, the composite number is bigger than 10^{15}, so for any supercomputer, this factorization challenge becomes nearly impossible to be solved in short amount of time. Even if somehow the computer manages to compute the factors, it is not necessary that for the next message, same composite numbers will be used.

So is this the only reason why RSA is still in use in security protocols and encryption systems? The short answer to this question is that since nobody knows how to compute the inverse RSA and without the prior knowledge of the prime factors of the composite number, nobody knows how to efficiently recover these prime factors from the composite itself. Hence, without knowledge of one key at least, this problem to decrypt the RSA is considered a hard polynomial problem.

With the development of quantum computers, the harnessed power of quantum mechanics promises to deliver exponential boost in the computational power. Considering the properties of quantum computers like superposition, wherein the possibility of a quantum bit being both 0 and 1 is non-zero, and entanglement, wherein two quantum bits share their data with each other, until measurement is performed [8]. The system can compute many solutions to the same problem in a single execution. According to the paper of Peter Shor [9] in 1994, quantum computers are powerful enough to break one of the best encryption algorithms today.

Shor's algorithm is composed of two parts: The first part of the algorithm turns the factorization problem into a problem of finding the period of a function that could be implemented over a classical computer. The second part finds the period using the Quantum Fourier Transform (QFT) and is responsible for the quantum speedup of the algorithm.

With the development of Shor's algorithm, that is able to compute the factors of large numbers of range 10^{10} in exponentially smaller time compared to classical computers, it became a crucial element in the process for cracking trapdoor-based codes. It is well understood that a few years ago, what is considered as the most secure and impossible to break algorithm, is now not only breakable but also could be cracked with much greater efficiency (see Figure 4.4).

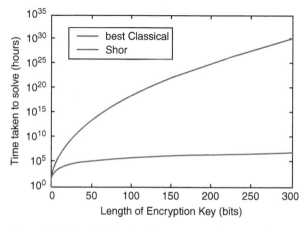

Figure 4.4 Graph indicating the quantum advantage in terms of computational boost. (*Source*: Quantum Computing/IBM).

This not only impacts the industries directly but many organizations working in security domain are feeling threatened already. Many scientists and developers working in this domain warn for bigger threats to classical encryption algorithms. It's no wonder that the quantum computers have the ability to crack even the world's strongest encryption algorithm in a short span of time. The US National Academies of Sciences, Engineering, and Medicine with their recent study on such algorithms recommends security organizations to buckle up and begin their development on super-powerful quantum algorithms to protect any crack in their conventional cryptographic defenses.

4.2 Quantum Key Distribution

A modern day problem always require a modern day solution. It is well known by now that quantum computers have the ability to break the most advanced locks of the digital system. But, is it also possible to create a lock entirely made up of a quantum computer? Quantum Key Distribution (QKD) is one such similar technique that is based on the principles of working on a quantum computer and enables two users to share a common secret key for cryptographic purposes [13].

Encryption is performed by combining the message with the key in such a way that the result is impossible to decode by an observer who have no knowledge of the key. Not to be confused about the purpose of QKD, that is to ensure the secrecy of the distributed key, and not to actually encrypt data. QKD requires a transmission channel (optical fiber) on that quantum carriers (photons) are transmitted from source to destination. In the quantum carriers, it is the source that encodes the pieces of information to generate the secret key. This is where QKD is particularly useful, as it can distribute as lengthy key as it is required by source user and destination user.

The security of QKD can be proven mathematically without imposing any restrictions on the abilities of a spy user monitoring on the channel. This behavior is unusual and it is something that was not possible with classical key distribution. This adds an unconditional security, although there are some minimal assumptions of quantum mechanics required such that the source user and destination user are able to authenticate each other and most importantly the spy on the channel should not be able to impersonate either as the source or destination user.

4.2.1 Advantage of Asymmetric Ciphers

A quantum computer poses great threat to current encryption algorithms. As a protective measure, there is an active effort by scientists to develop asymmetric ciphers that a quantum computer fails to defeat as explained above. While the potential utility of Shor's algorithm for cracking standard cryptography was a major driver of early enthusiasm in quantum computing research, the future development of quantum cryptographic algorithms are believed to be non-vulnerable and hack prone. This will reduce the usefulness of a quantum computer for cryptanalysis but from commercial point of view, this trade-off is settled with ease (see Figure 4.5).

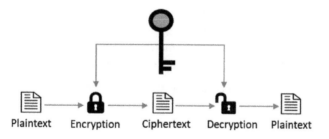

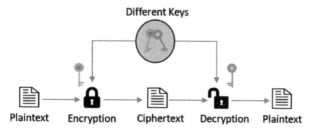

Figure 4.5 Difference between symmetric and asymmetric ciphering. (*Source*: Modified from Chandra et al., 2014 [14]).

4.2.2 The Security of QKD

All types of communication system are considered secure until some form of breach or taping occurs in the system. In early times, spies used to tap the telephone lines to listen to the conversation between two users in search of information. It was a problem at that time, hence the concept of superimposition of waves came handy to encrypt the conversation in a different form such that a spy user could not become a part of the communication channel or system.

If during the transmission of message between a source user and a destination user communicating over a quantum channel, suppose a spy user tries to be smart and might want to tap the channel in search for the potential secret key. Now what happens in quantum channel is that quantum bits flows through the channel that carries information from source to destination. Typically, a channel is optical fiber with photons acting as a carrier. Once the transmission is done, both intended users compare a segment of exchanged keys to identify if there are any transmission errors produced. So, if a source user sends a message to a destination, and the channel was noble and does not contain any spies monitoring, then only the message will be delivered to the destination and the comparison of keys will produce no errors. However, if a spy is monitoring the channel, then the property of quantum measurement, that is required by the spy to read the message, information collapses and it indicates both parties because while comparing the keys, error is generated due to unwanted measurement action done by the spy user.

4.2.3 Transmitting Data Using QKD

In classical computing, the data of any form whether it be a string, a number, or a character can be expressed as a binary set {0, 1}. In contrast, the quantum information cannot be expressed in classical terms, instead the unit of information is quantum qubits expressed in Dirac's notation as $|1\rangle$ or $|0\rangle$ or a superposition of both.

Let the source user A encode the random information (classical) bits using a set of 4 unique quantum qubits. The classical bit 0 can be encoded as either $|0\rangle$ or $|+\rangle = \left(\frac{1}{\sqrt{2}}|0\rangle + \frac{1}{\sqrt{2}}|1\rangle\right)$. The classical bit 1 can be encoded as either $|1\rangle$ or $|-\rangle = \left(\frac{1}{\sqrt{2}}|0\rangle - \frac{1}{\sqrt{2}}|1\rangle\right)$. Probability of A selecting any of the two basis are equally likely for encoding random information. The encoded data is transmitted to destination user B.

When photon approaches at B, it require to decode/decrypt the data to its original form. For this, B first of all requires to perform "measurement" of quantum qubits (see Table 4.1).

A measurement is a technique for quantum computers to determine the state of qubits. However, the adverse effect of this measurement results in collapse of the quantum state to either $|1\rangle$ or $|0\rangle$ and therefore a quantum state can no longer be in a superposition. We can say that it is a form of destructive transformation and so the destination user B cannot accurately determine the received qubits (if source user sent some information in $|+\rangle$ or $|-\rangle$ form).

Let us suppose, source user A sent all information encoded in $|1\rangle$ or $|0\rangle$ basis. Using measurement, destination user B can easily measure the data as $|1\rangle$ and $|0\rangle$ are orthogonal. However if A used $|+\rangle$ or $|-\rangle$ basis, destination user B is only able to read the collapsed information post measurement.

Assume $|\Phi\rangle$ as a superposition state determined as $\alpha|0\rangle + \beta|1\rangle$ where $|\alpha^2|$ and $|\beta^2|$ are the normalized probability coefficients of the two states with $|\alpha^2| + |\beta^2| = 1$. In the case when source user A had transmitted $|+\rangle$ or $|-\rangle$ to the destination user B, it will be able to measure $|1\rangle$ or $|0\rangle$ with a probability of $\frac{1}{2}$. Therefore B will not be able to distinguish the information and the information stands uncorrelated to the source user [7] (see Figure 4.6).

4.3 QKD Protocol: BB84

In 1984, Charles Bennett and Gilles Brassard [15] developed the first protocol with working action similar to public key cryptography. The implementation consists in the transmission from source user of 0° ($\updownarrow$) and 90° ($\leftrightarrows$) (the $|0\rangle$ and $|1\rangle$

Table 4.1 Transmission of classical bits as qubits in QKD.

Source User's Key	0	1	1	0	1	0	0	1
Source User's Basis	$|0/1\rangle$	$|+/-\rangle$	$|0/1\rangle$	$|0/1\rangle$	$|+/-\rangle$	$|0/1\rangle$	$|+/-\rangle$	$|0/1\rangle$
Polarized Photon Sent	$\updownarrow$	$\diagdown\diagdown$	$\leftrightarrows$	$\updownarrow$	$\diagdown\diagdown$	$\updownarrow$	$\diagup\diagup$	$\leftrightarrows$

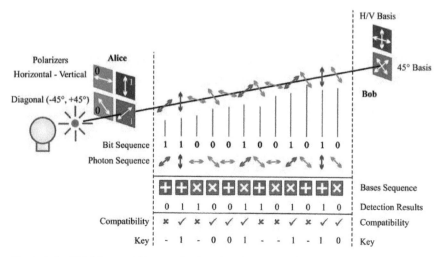

Figure 4.6 BB84 Protocol Communication showing usage of different bases and the resultant keys received at the recipient end to compare the original keys with the received keys.

basis), 45° (╱╱) and 135° (╲╲) (the |+⟩, and |−⟩ basis) using linear polarizations over optical fiber using a technique called polarization modulation. Source users select randomness in selecting the basis of transmission, information of that is reserved with the sender itself. At the receiving end, the four polarizations will usually appear changed, due to noise in the channel of communication, but before it can be analyzed the destination user is required to convert the same message into the original coordinate system by a suitable polarization controller. The destination user measures the information 50% time in |0⟩ and |1⟩ basis and the remaining 50% time in |+⟩ and |−⟩ basis. With this the source user also conveys the encoding schema it decides for each element of information to be encoded to the destination user.

This way, the destination user B is now able to reduce the wrong measurements that does not match with the schema of source user A and able to keep a good part of exact information as sent by the source user.

4.4 Evaluating the Case of Spying

If by some probability there is a scenario where a spy user S is monitoring the transfer of information between source user A and destination user B, we call the scenario eavesdropping. Assuming that none of the users have knowledge of this spy user S, the transfer is done using the above BB84 algorithm. However, as stated in an earlier section, to collect the information from the channel, a user has to perform the measurement. Once this spy user S proceeds with the measurement of the information, it actually is again destroying all the superposition in the channel. This is a fundamental property of quantum computing known as "no cloning," [16] that is first described by Wootters, Zurek, and Dieks in 1982, and has profound implications in quantum computing and related fields.

> *"The no cloning theorem is a result of quantum mechanics that forbids the creation of identical copies of an arbitrary unknown quantum state."*

Let $|\psi\rangle_A$ be the state of a quantum system A that a user desires to clone, therefore

$$|\psi\rangle_A = a|0\rangle_A + b|1\rangle_A. \tag{4.1}$$

The complex coefficients a and b can be any value. Similarly, there is a required identical hilbert space and initial state $|\varphi\rangle_B$ on that we desire to clone the state $|\psi\rangle_A$.

The composite system is then described by the tensor product, and its state is $|\psi\rangle_A \otimes |\varphi\rangle_B$. There are only two ways to manipulate the composite system, either we measure our observation thatthat irreversibly collapses the system into some eigenstate and corrupting the information contained in the qubit, or we could alter the hamiltonian of the system by using a time evolution operator $U(\Delta t)$, that is linear. For a fixed time interval Δt, again independent of $|\psi\rangle_A$, $U(\Delta t)$ will act as a copier if it produces no changes to the Hamiltonian of the system

$$U(\Delta t)|\psi\rangle_A|\varphi\rangle_B = |\psi\rangle_A|\varphi\rangle_B \tag{4.2}$$

where

$$|\psi\rangle_A|\varphi\rangle_B = \left(a|0\rangle_A + b|1\rangle_A\right)\left(a|0\rangle_B + b|1\rangle_B\right) \tag{4.3}$$

$$|\psi\rangle_A|\varphi\rangle_B = a^2|0\rangle_A|0\rangle_B + ab|0\rangle_A|1\rangle_B + ba|1\rangle_A|0\rangle_B + b^2|1\rangle_A|1\rangle_B \tag{4.4}$$

and similarly,

$$U(\Delta t)|\psi\rangle_A|\varphi\rangle_B = U(\Delta t)\left(a|0\rangle_A + b|1\rangle_A\right)|\varphi\rangle_B \tag{4.5}$$

must produce only results of pure states, i.e.,

$$U(t)|0\rangle_A|\varphi\rangle_B \approx a^2|0\rangle_A|0\rangle_B \tag{4.6}$$

and

$$U(t)|1\rangle_A|\varphi\rangle_B \approx b^2|1\rangle_A|1\rangle_B \tag{4.7}$$

therefore

$$U(\Delta t)|\psi\rangle_A|\varphi\rangle_B = a^2|0\rangle_A|0\rangle_B + b^2|1\rangle_A|1\rangle_B \tag{4.8}$$

And hence using (a) and (b) we can conclude that for all ψ

$$U(\Delta t)|\psi\rangle_A|\varphi\rangle_B \neq |\psi\rangle_A|\varphi\rangle_B \tag{4.9}$$

and hence $U(\Delta t)$ cannot act as a cloning aid. Hence cloning is not possible.

The problem with the spy user S will remains the same as it is for destination user B that it is also unable to identify the information sent in $|+\rangle$ and $|-\rangle$ basis and also disturb the original state of information. If the spy user S retransmit the state as detected, Alice and Bob will eventually capture the presence as the number of errors between their elements begins to increase.

4.4.1 Detecting an Eavesdropper

Optionally the user A and B can perform a test to detect the presence of a spy user on the channel. So far we know the intention of spy is to get the data communicated between source user A and destination user B and the challenges it might face while interpreting information from channel. As long as the spy user S continue to measure an element, it raises a probability of $\frac{1}{4}$ to introduce wrong information when sending the data back to destination user B.

Suppose an element sent from source user A is $|+\rangle$, when received by the spy user only have a probability 50% of measuring in the right basis. Suppose it does measure in correct basis, the original states will not be disturbed and the information propagates on the channel unnoticed. But as mentioned, the spy user will not be correct all the time. So if it does not measure in the correct basis, all S is doing is corrupting the original data and the state of element is disturbed. This disturbed state element contains information that is of no use and when this corrupt data is propagated to destination user B, all B is going to receive is just random information, that on measurement again is going to produce irrelevant information only (see Table 4.2).

We can say that there is only 50% chance that the elements of spy user is going to be correctly measured as intended by source user and remaining half time it does not. Since the data propagated is incorrect 50% time, then the intended users of information are able to interpret the presence of an unintended user by identifying too many errors during transmission.

Table 4.2 Quantum error caused while eavesdropping.

Source user's key	0	1	1	0	1	0	0	1
Source user's basis of transmission	+	X	+	+	X	+	X	+
Polarized photon sent	↕	↘↖	⇆	↕	↘↖	↕	↗↗	⇆
Spy user's measurement basis	+	+	X	+	+	X	X	X
Polarized photons measured by spy user	↕	⇆	↘↖	↕	⇆	↗↗	↗↗	↘↖
Destination user's measurement basis	+	X	+	X	+	X	+	X
Polarized photons measured by receiver	↕	↘↖	⇆	↗↗	⇆	↗↗	↕	↘↖
Public discussion of basis	+	+	+	+	+	+	+	+
Error in keys	✓	✗	✓	✗	✗	✗	✗	✗

4.4.2 Measures to Prevent Eavesdropping

Upon detecting too many errors, the source and destination user might abort the protocol, the reason being eavesdropping. However, in practical scenarios, the physical implementation is not 100% perfect and the errors might arise due to noise in the channel, or quantum losses, or imperfect quantum states generation or detector imperfection.

Now to overcome this error, both source and destination users have to count the occurrence of errors occurred in the disclosed key elements. The count is then divided by the number of disclosed key elements n to obtain a fraction of transmission errors in set of disclosed key elements.

$$e = \frac{\text{no. of errors in disclosed key elements}}{\text{total disclosed key elements}} \tag{4.10}$$

This fraction is termed as a bit error rate and it can be used to determine the amount of information a spy user gained from the key elements.

But the spy user remains unclear about the secret key. Since source and destination users now have information on how much error is in the transmission e, and only they know the amount of actual length of undisclosed secret key l and disclosed length n. Using two more steps of a process called Secret-Key Distillation (SKD), the source and destination user can then gain the full secret key using an authentic public channel.

4.4.3 Error Correction

The SKD process consists of two steps: reconciliation and privacy amplification.

The process of reconciliation is sort of an interactive error correction protocol where the source user A and destination user B disclose parts of their key elements in an alternate manner. If there is an odd number of errors in the corresponding subset, using a procedure called dichotomy, the error location can be identified and procedures to correct it can be implemented.

The source user and destination users, on gaining knowledge of this spy user are required to exploit the information this spy user gained in order to produce confusion for the spy user such that it becomes unable to gain secret key as well as for intended users to obtain the secret key. This is the idea behind privacy amplification where the exploitation of information gained by spy user is done. The source user and destination user are required to deduce a function f like a hash function for their key elements using that the confusion of the spy user amplifies for the entire information it gained. As we learned that for SKD, the transfer of information is required to be made over the public authenticated classical channel, it is assumed that there is no need of any calibration for the channel. During reconciliation, both the source user and the destination user need to agree on the same hash function they are preparing to use. The new information after processing it through a hash function is transmitted over the channel that is eavesdropped. The technique is termed as asymmetric ciphering.

We can say that now the knowledge of this spy user is composed of two things, the first is the information that it gained earlier from channel between source user and destination user, call it *I* bits and the second obtained from the process of reconciliation where new information passed, call it *K* bits. The spy user had information of *I* bit but is completely unaware of *K* bits as it does not know what to perform with these bits.

If the function *f* produces an output of $I + K \bmod 2$, the spy user will have literally no clue on what to extract from the information as the two possibilities of $I + K \bmod 2 = 0$ and $I + K \bmod 2 = 1$ are equally likely irrespective of the value of *I*.

The work on privacy amplification is done in such a manner that the completely secret output key remains smaller than the partially disclosed input secret key. The difference in the size of the keys is equal to the number of bits known to the spy user. To prohibit the spy user to gain any knowledge about the secret key, the reconciliation process is required to disclose as small knowledge on the key as possible, but large enough such that the source and destination users are able to correct all their errors. At last, the secret key obtained after privacy amplification procedure can be declared useful as the source user and destination user can use it for cryptographic purposes to encrypt messages over the secret channel free from eavesdroppers.

4.5 Challenges in Quantum Computing

The QKD has the advantage of being future-proof unlike classical key distribution, and it is definitely not possible for a user-friendly protocol for spy user to trace a quantum signals, owing to the quantum non-cloning theorem. However, the challenge is to maintain the polarization of photons. While traveling through the channel, the polarization of photon may change due to external noise or interferences or environmental causes that corrupt the message. The quantum signals are virtually hack-proof, but still originality of the message could be damaged by the spy user that is a big challenge yet to be solved. Another challenge is the implementation of quantum crypto system, that will require a very costly infrastructure to develop, maintain, and work with. Another limitation comes with the length bound to set up a network of optical cables, couplers, and optical amplifiers [17]. Development and working with a quantum computer require expertise not only in device engineering but also in most scientific disciplines from computer science and mathematics to physics, chemistry, and materials science, and finding a candidate with such intellect is another major challenge.

4.6 Conclusion

In the chapter we had studied in depth about the theory of cryptography, how basic cryptography is done and the varieties of cryptography. Further we learned about the advantages of cryptography and introduced the quantum algorithms for performing the same cryptography. Further we learned about generation of quantum keys and its benefits toward security in QKD and use of the same in the BB84 protocol is explained. The process of encrypted communication is also explained with a demonstration of how a message is transmitted from source to destination with the use of quantum channels. Noise creates a hinder effect in any form of communication and so does quantum cryptography, that we learned and also the methodologies to prevent any noise to corrupt the information. The noise or errors are unavoidable, still it could be corrected using error correction methodologies through that end users can recover the original intended message from the corrupted data. Lastly, we understood the challenges in quantum computing and practical challenges that might obstruct the working process of such technologies in present day scenario.

Bibliography

1 Britannica Encyclopaedia. (July 2020). Cipher. https://www.britannica.com/topic/cipher accessed June 19, 2021.

2 Wobst, R. (2001). *Cryptology Unlocked*, 19. Wiley. ISBN:978-0-470-06064-3.

3 Luciano, D. and Prichett, G. (January 1987). Cryptology: from Caesar ciphers to public-key cryptosystems. *The College Mathematics Journal* 18 (1): 2–17. doi: 10.2307/2686311.

4 Bellare, M. and Rogaway, P. (September 2005). *Introduction to Modern Cryptography*, 10–80. UC DAVIS.

5 Dentzel, Z. How the internet has changed everyday life. BBVAOpenMind. https://www.bbvaopenmind.com/en/articles/internet-changed-everyday-life, accessed June 19, 2021.

6 Diffie and Hellman. (November 1976). New directions in cryptography. *IEEE Transactions on Information Theory* 644–654. doi: 10.1109/TIT.1976.1055638, ISSN:0018-9448.

7 Stallings, W. (May 1990). *Cryptography and Network Security: Principles and Practice*, 7–165. Prentice Hall. ISBN:9780138690175.

8 Nielsen, M. and Chuang, I. (2010). *Quantum Computation and Quantum Information: 10th Anniversary Edition*. Cambridge University Press. doi: 10.1017/CBO9780511976667.

9 Shor, P. W. (January 1996). Polynomial-time algorithms for prime factorization and discrete logarithms on a quantum computer. doi: 10.1137/S0097539795293172.

10 Compton, A. H. and Heisenberg, W. (1984). *The Physical Principles of the Quantum Theory*, 117–166. Scientific Review Papers, Springer. doi: 10.1007/978-3-642-61742-3_10, ISSN:978-3-642-61742-3.

11 Dirac, P. A. M. (1958). *The Principles of Quantum Mechanics*, 4e. Oxford University Press, Oxford. ISBN:0-19-851208-2.

12 Rivest, R., Shamir, A., and Adleman, L. (February 1978). A method for obtaining digital signatures and public-key cryptosystems. *Communications of the ACM* 21 (2): 120–126. doi: 10.1145/359340.359342.

13 Bennett, C., Brassard, G., and Ekert, A. (1992). Quantum cryptography. *Scientific American* 267 (4): 50–57.

14 Chandra, S., Paira, S., Alam, S. S., and Sanyal, G. (2014). A comparative survey of symmetric and asymmetric key cryptography. *International Conference on Electronics, Communication and Computational Engineering* 83–93. doi: 10.1109/ICECCE.2014.7086640.

15 Bennett, C. H. and Brassard, G. (1984). Quantum cryptography: public key distribution and coin tossing. *IEEE International Conference on Computers, Systems and Signal Processing* 175: 8.

16 Wootters, W. and Zurek, W. (1982). A single quantum cannot be cloned. *Nature* 299 (5886): 802–803. doi: 10.1038/299802a0.

17 Grzywak, A. and Kowalczyk, G. P. (2009). *Quantum Cryptography: Opportunities and Challenges*, 195–215. Springer. doi: 10.1007/978-3-642-05019-0_22, ISSN:978-3-642-05019-0.

5

Quantum Machine Learning Algorithms

Renata Wong[1], Tanya Garg[2], Ritu Thombre[3], Alberto Maldonado Romo[4], Niranjan PN[5], Pinaki Sen[6], Mandeep Kaur Saggi[7] and Amandeep Singh Bhatia[8]

[1]*Physics Division, National Center for Theoretical Sciences, Taipei, Taiwan ROC*
[2]*Indian Institute of Technology, Roorkee, Uttarakhand, India*
[3]*Computer Science and Engineering Department, Visvesvaraya National Institute of Technology, Nagpur, India*
[4]*Centro de Investigación en Computación, Instituto Politécnico Nacional, Mexico City, Mexico*
[5]*Amrita Vishwa Vidyapeetham, Coimbatore, Tamil Nadu, India*
[6]*National Institute of Technology, Agartala, Tripura, India*
[7]*Computer Science and Engineering Department, Thapar Institute of Engineering & Technology, Patiala, Punjab, India*
[8]*School of Electrical and Computer Engineering, Purdue University, West Lafayette, IN, USA*

5.1 Introduction

The artificial intelligence and machine learning are important tools for solving real-world problems and have got significant attention among scientific research community, business and industry. In the past decade, machine learning traversed a vast array of algorithms ranging from speech recognition, self-driving cars, web searching effectively, and enormously improved the realization of the human genome. The machine learning algorithms learn by an interplay and upgrade automatically via experience using data [1, 2]. The algorithms are divided into mainly three categories i.e., supervised learning for task-driven tasks, unsupervised learning for data-driven tasks and reinforcement learning to master from mistakes [3]. The machine learning finds applications in various fields such as robotic control, prediction of weather, span mail filters, alerts for traffic, risk assessment in finance, fraud detection, image/text recognition, social media, medical diagnosis and many more [4, 5]. It involves an emergence of massive datasets of input-output pairs from new sources called big data. The objective is to process the data effectively and efficiently by applying machine learning algorithms.

The field of quantum computing has the potential to revolutionize other scientific areas because of quantum mechanics principles. In quantum computing, factorization of integers is exponentially faster and unordered search in the database is quadratically faster than existing classical algorithms [6–8]. It is known to solve the tasks which cannot be tractable using a classical computer. Several quantum learning algorithms depend on the application of Grover's searching algorithm [9]. There are four methods that describes how the machine learning and quantum computing can be combined [10]. It relies upon whether the information processing is performed by a classical (C) or quantum (Q) device and data generated by a classical or quantum device, as shown in Figure 5.1.

The term CC refers how classical data being processed by classical computers. The approach QC demonstrates how classical machine learning can be help with quantum computing [11]. In this chapter, we focus on the CQ method and how classical datasets can be processed with quantum computers, called quantum

Emerging Computing Paradigms: Principles, Advances and Applications, First Edition.
Edited by Umang Singh, San Murugesan and Ashish Seth.

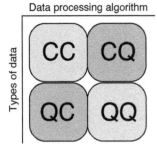

Data processing algorithm

Types of data

Figure 5.1 The four methods to integrate machine learning and quantum computing.

machine learning. Till now, very few results exist for QQ method i.e., how quantum data generated by quantum device can be processed by quantum computers. Quantum machine learning have shown great potential and provide a wide array of applications for small size quantum systems, quantum information and digital processors [12]. Recently, the combination of quantum computing and tensor networks is getting a significant attention and proposed several quantum machine learning algorithms for different tasks [13–17].

5.2 Quantum Machine Learning

Quantum machine learning is developed based on mainly three schemes:

1) Harrow, Hassidim, and Lloyd (HHL) algorithm to invert linear system of equations by using quantum mechanics [18, 50].
2) Grover's searching algorithm, which is used for the amplitude amplification to uncover an unstructured search problem and gain quadratic speedup e.g., quantum variant of k-nearest neighbours [19], k-mean and k-median algorithms [20, 21] are based on Grover's searching algorithm.
3) Recently, some implementation has been performed based on quantum annealing to classify and binding proteins in a gene [22].

There are nine well-known quantum machine learning algorithms, as shown in Table 5.1:

1) Quantum k-nearest neighbor method [23]
2) Quantum principal component analysis [21]
3) Quantum support vector machine [24, 18]
4) Quantum reinforcement learning (Dong et al., 2008)[51]
5) Quantum Boltzmann machines [25]
6) Quantum clustering (Aimeur et al., 2007)[52]

Table 5.1 The outline of quantum machine learning algorithms.

Quantum machine learning algorithms	Speedup	Quantum data	CC	CQ	QC	QQ
Quantum k-nearest neighbor method [23]	Quadratic	No			⁎	
Quantum principal component analysis [21]	Exponential	Yes	⁎			
Quantum support vector machine [24, 18]	Quadratic	No			⁎	
	Exponential	Yes				
Quantum reinforcement learning (Dong et al., 2008)	-	Yes			⁎	

(Continued)

Table 5.1 (Continued)

Quantum machine learning algorithms	Speedup	Quantum data	CC	CQ	QC	QQ
Quantum Boltzmann machines [25]	Exponential	No			⊹	
Quantum clustering (Aimeur et al., 2007)	Quadratic	No		⊹		
Quantum neural networks	Exponential	Yes			⊹	
Quantum generative adversarial networks [26]	Exponential	Yes			⊹	
Quantum autoencoder [Khoshaman et al., 2017, 27]	Exponential	Yes			⊹	

7) Quantum neural networks
8) Quantum generative adversarial learning [26]
9) Quantum autoencoder [Khoshaman et al., 2017, 27]

Let's examine them in detail.

5.2.1 Quantum *k*-nearest Neighbor Algorithm

In supervised learning, the k-NN (*k*-nearest neighbors) algorithm is a well-known classification algorithm. It is called k-NNas it classifies from a test set states $|u_i>$ whose labels are to be determined, in order to find the minimum distance to the states of the train set $|v_i>$, whose labels are known to us. In order to find the distance between two instances, the Euclidean distance is used in classical computing,

$$|u - v| = \sqrt{\sum Ni = 1(ui - vi)^2} \qquad (5.1)$$

For quantum computing, it is achieved by the inner product (or dot product) which is defined as

$$|u - v| = |u||v| - u \cdot v, \qquad (5.2)$$

where u and v are two quantum states [19]. In order to implement Equation (5.2), it is necessary to use a quantum subroutine that can estimate the distance between two quantum states. This is possible from the SWAP test, which is a quantum algorithm that can be used to estimate the fidelity of two pure states |u> and |v>, i.e., $F=|<u|v>|^2$. This can be generated from three quantum registers, two to represent the instances to be compared |u>,|v> and one from a state |0> ancilla which will be obtained as the result of such evaluation,

$$|0\rangle \otimes |u\rangle \otimes |v\rangle \qquad (5.3)$$

Subsequently, the circuit is designed to generate such an operation with a Hadamard gate in the first qubit (ancilla) followed by a controlled SWAP in the other two registers where the first qubit serves as the control system, after add another Hadamard gate and measurement the value of ancilla, as shown in Figure 5.3.

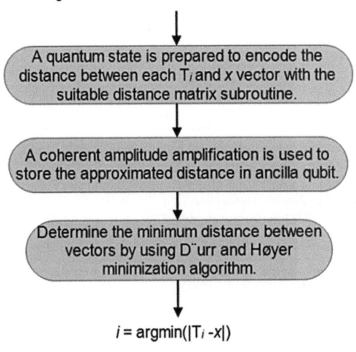

Figure 5.2 The operation of quantum k-NN algorithm.

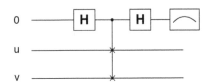

Figure 5.3 The quantum circuit of SWAP test.

Applying the Hadamard gate to the first qubit that is in the $|0>$ state is to pass it to the $|1>$ state as follows

$$\frac{1}{\sqrt{2}}\left(|0\rangle|u\rangle|v\rangle+|1\rangle|u\rangle|v\rangle\right), \tag{5.4}$$

The CSWAP gate is composed so that when the control qubit is set to $|0>$ the states are maintained, otherwise it is set to $|1>$ the states are exchanged, leaving the expression

$$\frac{1}{\sqrt{2}}\left(|0\rangle|u\rangle|v\rangle+|1\rangle|v\rangle|u\rangle\right), \tag{5.5}$$

Followed by a second Hadamard gate in the first qubit is as follows

$$\frac{1}{2}\Big[|0\rangle\big(|u\rangle|v\rangle+|u\rangle|v\rangle\big)+|1\rangle\big(|v\rangle|u\rangle-|u\rangle|v\rangle\big)\Big],\tag{5.6}$$

Finally, the measurement is performed on the first qubit, we obtain the following probabilities

$$\mathrm{P}(0)=\frac{1}{2}+\frac{1}{2}\langle u|v\rangle^{2}\tag{5.7}$$

$$\mathrm{P}(1)=\frac{1}{2}-\frac{1}{2}\langle u|v\rangle^{2}\tag{5.8}$$

Considering how to generate a distance between two quantum states, one has to decide whether the following quality is true:

$$min\big|x-x_{j}\big|\leq min\big|x-y_{j}\big|;\ x_{j}\in C1,\ y_{j}\in C2\tag{5.9}$$

where $C1$ and $C2$ are the states defining the classes. There are some considerations for this quantum algorithm [28–30].

- The quantum random access memory (qRAM) structure can be used to search the entire database of quantum states with $O(\log 2^{n})$ complexity.
- The result is the estimation of unknown quantum states, by estimating their parameters.
- It therefore has the ability to classify unknown states.
- It is applicable to solve the handwritten recognition and traveling salesman problem.

5.2.2 Quantum Principal Component Analysis

The principal component analysis (PCA) algorithm is used to lower the data dimensions. PCA is a technique used in dimensionality reduction [31, 32]. Given a vector in m-dimensional Hilbert space, PCA maps it to n-dimensional subspace $(n<m)$. Original vector is mapped on to n directions, which are the principal components, and span the subspace. The simplest way of finding PCA is by determining the projections that results in increasing the variance. The direction in space is the first principal component along which the projections show maximized variance. The direction which maximizes the variance among all directions orthogonal to the first is the second principal component. Hence, the kth principal component is the variance orthogonal to the previous $k-1$ components results in maximizng the variance.

Consider m dimensional vector $\vec{v}=[v_{1},v_{2},\ldots,v_{m}]^{T}$. Now, to find PCA of $\vec{v}$ in n dimensions $(n<m)$, we first calculate the convarience matrix of $\vec{v}$ as shown in Equation (5.10).

$$C=\sum_{j=1}^{m}v_{j}v_{j}^{T}\tag{5.10}$$

Then, we computed C, where $\vec{c}_k$ is the eigenvector of C and e_k is the corresponding eigenvalue.

$$C = \sum_{k=1}^{m} e_k c_K c_k^T \tag{5.11}$$

Finally, PCA of $\vec{v}$ are the top n largest values of e_k. Complexity of classical algorithm for computing PCA is $O(d^2)$, where d is the dimension of the Hilbert space.

Lloyd et al. [33] introduced the quantum version of principal component analysis (QPCA) using Hamiltonian simulation and phase estimation. The proposed QPCA is exponentially superior to classical PCA. QPCA exploits the capability of Quantum Phase Estimation (QPE) algorithm to calculate eigenvalue of an unknown unitary matrix U, for a certain quantum state $|\psi\rangle$ as shown in Equation (5.12).

$$U|\psi\rangle = k|\psi\rangle \tag{5.12}$$

In order to find PCA of a vector $\vec{v} = [v_1, v_2, \dots, v_m]^T$, we convert $\vec{v}$ to a quantum state v. First, subtract mean $\bar{v}$ of $\vec{v}$ from its components Equation (5.13) and then normalize $\vec{v}$, as shown in Equation (5.14) [34, 35].

$$v_j = v_j - \bar{v}, \bar{v} = \frac{1}{m}\sum_{j-1}^{m} v_j \tag{5.13}$$

$$v_j = \frac{v_j}{|v|}, |v| = \sqrt{\sum_{j-1}^{m} v_j^2} \tag{5.14}$$

$$|v\rangle = \sum_{j=1}^{m} v_j j \tag{5.15}$$

Now, we obtain density matrix σ of this state $|v\rangle$ as shown in Equation (5.16).

$$\sigma = \frac{1}{m}\sum_{j=1}^{m} |v_j\rangle\langle v_j| \tag{5.16}$$

After calculating density matrix σ, QPE uses conditional application of $U = e^{-i\sigma t}$ for varying times t on an initial state $|\psi\rangle|0\rangle$ and maps it to $\sum \psi_i |\chi_i\rangle|r_i\rangle$, where χ_i are the eigenvectors of σ and r_i are estimates of the corresponding eigenvalues. Using σ itself as the initial state, QPE yields Equation (5.17).

$$\sum_i r_i |\chi_i\rangle\langle\chi_i| \otimes |r_i\rangle\langle r_i| \tag{5.17}$$

Sampling from this state allows us to reveal features of the eigenvectors and eigenvalues of σ. Eigenvalues of σ can now be used to calculate PCA of $\vec{v}$. Complexity of quantum algorithm to find PCA is $O(log(d)^2)$ [32].

5.2.3 Quantum Support Vector Machine

Quantum Support Vector Machine (QSVM) is the quantum analogous algorithm of the classical Support Vector Machine (SVM) algorithm [18]. SVM is a supervised machine learning algorithm that is used for binary classification of data. It can perform nonlinear classification by using functions to map the data to a higher dimensional feature space. These functions are called feature maps. The obtained higher dimensional feature space allows the construction of a hyperplane separating the two classes of data. The hyperplane is determined by measuring its distance from the nearest training data point of any class and finding the maxima to minimize the classification error. These data points are called the support vectors. The mathematical formulation of the optimization problem involves taking the inner products of each pair of data point vectors in the higher dimensional space. This collection of inner products is called kernel given by $K\left(x_i,x_j\right)=\overrightarrow{x_i}\cdot\overrightarrow{x_j}\ i,j=1,2,3\ldots\ldots N$ where $\overrightarrow{x_i}$ is a data point vector.

QSVM is a quantum machine learning algorithm that classifies classical data using quantum logic [36]. Using the QSVM algorithm is advantageous as it allows us to classify a bigger dataset with a higher speed and efficiency. This increase in speed is achieved by using quantum logic and its properties of entanglement and parallel computing. QSVM indicates an exponential speedup over the classical algorithm i.e., O(log(NM)) where N is number of features and M denotes the number of training points. The original, enlarged feature space and nonlinear decision boundaries are represented in Figure 5.4.

As the name of the algorithm suggests, SVM and QSVM represents its data points as vectors and have a hyperplane that can be represented by a vector [24]. In SVM the hyperplane vector is given by $\sum_i \alpha_i K\left(x_i,x\right)=constant$, which in the vector notation can be written as

$$\vec{w}\cdot\vec{x}+b=0$$

where $\vec{w}$ is the normal vector from the data-point to the hyperplane. The $\vec{w}$ is given by $\vec{w}=\sum_{i=1}^{N}\alpha_i\overrightarrow{x_i}$. To perform the quantum classification the following steps are performed in the given order:

- *Classical pre-processing*: Before feeding the classical data to the quantum computer, one needs to perform various pre-processing processes to extract the required data and features from a larger dataset and make the computations more analytically feasible. These processes include scaling, normalization and dimensionality reduction. The reformed classical data is then fed to the quantum machine, which will perform quantum transformations on the converted classical data to create the hyperplane. This process is done with the help of quantum feature maps.

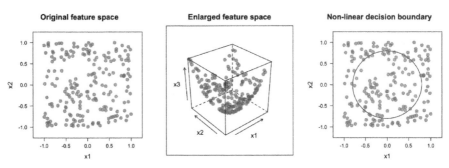

Figure 5.4 The schematic representation of the original feature space, enlarged feature space, and nonlinear decision boundary.

- *Quantum encoding of data*: The initial quantum state given by $|\chi\rangle = 1/\sqrt{N_x} \sum_{i=1}^{2^n} |\overrightarrow{x_i}| |i\rangle |x^i\rangle$ with $N_x = \sum_{i=1}^{2^n} |\overrightarrow{x_i}|^2$ can be constructed by using the qRAM oracle. These $|x^i\rangle$ states are the 2^n dimensional basis of the training vector space so that every training vector can be represented as a superposition of the basis vectors $|x_i\rangle = \sum_i \alpha_i |x^i\rangle$. To create a higher dimensional feature space that represents the features more concretely feature maps are used. A feature map is a function where F is the feature space. A quantum feature map is a feature map where F is a Hilbert space and the feature vectors are quantum states. It is a set of unitary transformations $U(x)$, which map $|x\rangle \rightarrow \phi(x)$. These transformations and thus the parameters of $U(x)$ depend on the data we are trying to classify. Thus, a feature map helps in dimensionality reduction as it selects certain features of the training data and reduces the number of variables to what is required to classify the selected features only [53]. A number of quantum packages like Qiskit have in-built feature maps like ZFeatureMap, ZZFeatureMap, and the PauliFeatureMap. For QSVM the kernel is defined as $K(x_i, x_j) = \phi(\overrightarrow{x_i}) \cdot \phi(\overrightarrow{x_j})$.

- *Construction of the hyperplane*: The hyperplane is determined by the kernel method, which is the least square problem: $\begin{pmatrix} 0 & 1^T \\ 1 & K + \gamma^{-1}I \end{pmatrix} \begin{pmatrix} b \\ \vec{\alpha} \end{pmatrix} = \begin{pmatrix} 0 \\ \vec{y} \end{pmatrix}$.

  To find the value of b and we need to find the inverse of the matrix $F = \begin{pmatrix} 0 & 1^T \\ 1 & K + \gamma^{-1}I \end{pmatrix}$ that can be easily done by splitting the F matrix into $F = J + K$ where $J = \begin{pmatrix} 0 & 1^T \\ 1 & 0 \end{pmatrix}$. The quantum state $\begin{pmatrix} 0 \\ \vec{y} \end{pmatrix}$ is given by $1/\sqrt{N_y} \left(|0\rangle + \sum_{j=1}^{M} y_j |j\rangle \right)$. The final state after taking the inverse and applying the inverted matrix on the above quantum state is

  $$\begin{pmatrix} b \\ \vec{\alpha} \end{pmatrix} = \frac{1}{\sqrt{N_y} \begin{pmatrix} b|0\rangle \\ \sum_{j=1}^{M} \alpha_j |j\rangle \end{pmatrix}}.$$

- *Optimizing the hyperplane equation*: The optimization process by using classical optimizers like COBYLA, SPSA, and SLSQP. These optimizers reduce the cost function value to minimize the value and maximize the distance from the nearest datapoints. After the optimization process, the final classification is done by the equation:

  $$y(\overrightarrow{x_0}) = \text{sgn}\left[\sum_{j=1}^{M} a_j (\overrightarrow{x_0} \cdot \overrightarrow{x_0}) + b \right].$$

  The signum function returns a 1 or -1 depending on the class, which the datapoint $\overrightarrow{x_0}$ belongs to. If the model learned ovefits the data, then the performance is poor. It is crucial to determine how QSVM performs on real-world datasets in future.

5.2.4 Quantum Boltzmann Machine

Boltzmann machine (BM) is a type of recurrent neural network. The neurons produce stochastic decisions regardless of true or false. The major goal is learn the actions of visible neurons and uncover some engaging properties. The solution of a problem is optimized. BM is a type of probabilistic machine learning algorithm that makes use of the Boltzmann distribution. It comprises a probabilistic network of binary units with a quadratic energy function. These BM are used as recommendation systems. Example: To recommend movies or products in online stores. There are various types of BM machines and can be divided into three types, BM (Figure 5.5: leftmost), restricted BM (middle), and deep BM (rightmost).

The basic structure of the BM is two types of nodes: visible and hidden. Each of these nodes is connected in a different fashion depending upon the type of BM used. These nodes can have values between 0 to 1 (classical algorithm) and each connection is associated with a weight. For the sake of this chapter let's name each node as z_i, each weight as w_i and a bias b_i. Thus, an energy function can be defined as Ising model with given configuration as

$$E = -\sum_a b_a z_a - \sum_{a,b} w_{ab} z_a z_b$$

This defined energy function can be considered as the loss function. The dimensionless parameters b_a and w_{ab} are tuned as such that the energy E is minimum. At the equilibrium (lowest energy) the probability of observing a state v of the visible variables is given by Boltzmann distribution summed over the hidden variables as

$$P_v = Z^{-1} \sum_h e^{-E_z}$$

where $z = \sum_z e^{-E_z}$. At first, sight implementing this classical with neural networks seems easy and feasible. But with an increase in the number of nodes or input data (visible nodes) the complexity becomes high and there requires a quantum computer such that one can exploit principles of quantum superposition and entanglement [25]. The energy function defined for the BM is an Ising Hamiltonian that can be efficiently minimized using quantum annealing devices such as DWAVE 2000 because the minimizing Ising Hamiltonian is a type of quadratic unconstrained

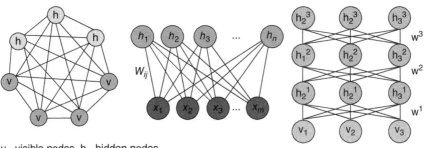

v - visible nodes, h - hidden nodes

Figure 5.5 The representation of BM, restricted BM, and deep BM.

binary optimization (QUBO) problem. A general objective function for QUBO as a form

$$E\left(a_i, b_{ij}, q_i\right) = \sum_i a_i q_i + \sum_{i<j} b_{i,j} q_i q_j.$$

It can be understood to be the same form of the energy function of the BM. One change that is made in the quantum BM over the classical BM is in QBM the nodes while classically $z_i \in \{0,1\}$. The main advantage over classically running this algorithm is in using quantum annealing one can theoretically reach global minima thus extracting a maximum accuracy [53]. Also, one can harness the power of quantum superposition to input large numbers of data. The QBM can be used as a monitoring device in nuclear power plants. Let us consider a nuclear power plant with all of its dial being binary. So one gets a set of binary numbers about the state of the power station. The task is to notice an unusual state. One way it can be done is to train a supervised machine learning model and deploy that as a monitoring device, but we don't want any examples of nuclear power station disaster. So the intelligent move will be using a BM with all the dials as visible layers along with some hidden layers [37]. After optimizing the energy function E, the equilibrium state E_0 corresponds to the correct functioning of nuclear power plants. Thus, whenever there is a fault in any of the nuclear power plants, the energy of the BM will be a greater E_0 thus this setup can be used as a monitoring device.

5.2.5 Quantum Neural Networks

In this section, we will assume a general understanding of classical neural networks. As a brief reminder, an example neural network is included in Figure 5.6. A classical neural network consists of three major layers of neurons (also called perceptrons). The input layer contains the input values, which in Figure 5.6 are x_1, x_2, and x_3. Layers 1 and 2 are called *hidden layers* and are used to train the network by means of weights, which can be different for each pair of perceptrons in two adjacent layers. The last layer is the output layer, here consisting of perceptrons y_1 and y_2.

The structure is a neural network corresponds to a function $f\left(x, \theta\right) = \sigma\left(wx + b\right)$, where w are the weights, x are the inputs, b are the biases of perceptrons, and σ is the *activation function* whose purpose is to "activate" certain perceptrons in the following layer. In order to train the model, we need to define a cost function

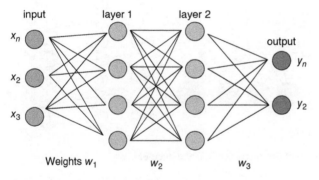

Figure 5.6 An example classical neural network.

$C = \sum_{x,y} |f(x, \theta) - y|^2$ that will let us know how good our parameters w and b are by comparing the actual output with the expected output. The motivation for developing a quantum model of neural networks was the fact that classical neural networks are difficult for large data training tasks. Besides being able to take care of this problem, it is believed that quantum neural networks could provide a certain speedup over their classical counterparts as well as model scalability [38].

The central idea behind quantum neural networks is that quantum circuits can be trained the same way classical neural networks are trained. To that end, we need to parameterize quantum circuits so as to be able to carry out optimization procedures for our cost function. In this section, we will focus on one such procedure that is *gradient descent*. Parameterized quantum circuits are also referred to as *variational quantum circuits*. An example of such a variational circuit is given in Figure 5.7.

A variational quantum circuit consists of the following general steps:

1) Preparation of some initial state $|\psi\rangle$, usually initialized to $|0\rangle^{\otimes n}$ where n is the number of qubits.
   In Figure 5.7, initialization involves three rotation gates with varying angles. For example, $Ry\left(-\dfrac{\pi}{2}\right)$ is a clockwise rotation about the y axis by 90 degrees, while $Rz\left(\dfrac{\pi}{4}\right)$ is a counterclockwise rotation about the z axis by 45 degrees. Positive angles refer to a counterclockwise rotation, while negative angles refer to a clockwise rotation.
2) Execution of a series of parameterized unitary transformations $U(\theta)$. These transformations correspond to a series of quantum gates.
   In Figure 5.7, there are four unspecified parameters: θ_1, θ_2, θ_3 and θ_4. The values of these parameters will depend on the gradient values obtained for a particular cost function.
3) Measurement of a subset of the qubits as output. Usually the measurement is carried out in the Pauli Z basis with eigenstates $|0\rangle$ and $|1\rangle$. In Figure 5.7, the last two gates are measurements on the top and the middle qubits, respectively. These outputs give us the cost function for the circuit.

Given a variational quantum circuit, we can train it by adjusting the parameterization θ and feeding it back to the circuit after the gradient has been computed. Fig. shows a schema that illustrates the process. Here, θ is a parameterization for the

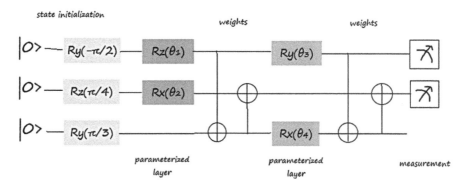

Figure 5.7 An example of a variational quantum circuit.

quantum gates in a circuit. The circuit in the example consists of three rotation gates corresponding to rotations around the *x*, *y*, and *z* axis, respectively, by an angle θ. is the cost function for the circuit, while ∇f is the gradient of the circuit's output with respect to the circuit's parameters θ. Based on the gradient descent, the modified parameters θ are then fed back to the circuit and the circuit is executed again.

The cost function can be seen as the expectation value of circuit's outputs and depends on the problem we want to solve. The purpose of model optimization is to minimize the expectation value. To that end, we can use e.g., gradient descent. Here we will focus on one of the methods for computing gradients (which are vectors of partial derivatives with respect to a parameterization θ), which is the parameter-shift rule. Figure 5.8 represents the schema for network training by gradient calculation.

The parameter-shift rule is deduced from the fact that if the function $f(\theta) = \sin(\theta)$ can be evaluated, then its first partial derivative with respect to the angle θ can be directly obtained by the following formula:

$$\frac{\partial f(\theta)}{\partial \theta} = \frac{f\left(\theta + \frac{\pi}{2}\right) - f\left(\theta - \frac{\pi}{2}\right)}{2}.$$

In this formula, the derivative is calculated by shifting the value of by $\frac{\pi}{2}$ forwards and backwards and then subtracting the second value from the first. For sinusoidal functions, this formula can be abstracted to a parameter as follows:

$$\frac{\partial f(\theta)}{\partial \theta} = s\left(f(\theta + p) - f(\theta - p)\right) = f(\theta_1) - f(\theta_2).$$

We observe that the above formula requires at least two executions of a circuit, each for one of the two shifts, in order to compute the respective gradient value. As gradients are vectors of partial derivatives, a circuit of *n* parameters would require $2n$ evaluations.

Note that all single qubit gates can be realized as rotations in the complex Hilbert space, that allows us to use the above formula. This in general does not apply to two-qubit gates however. For these gates various other methods must be applied, such as approximation methods.

An important feature of the parameter-shift formula is that it can be readily extended to calculate higher-order derivatives, such as Hessians, for the purpose of cost function optimization. Furthermore, the previously obtained derivatives can often be reused in the following iteration of the optimization process [39].

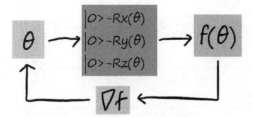

Figure 5.8 Schema for network training by gradient calculation and feedback into quantum circuit.

Consider an example of the quantum circuit given in Figure 5.7. It is computed as follows (for the sake of simplicity, we omit the CNOT gates corresponding to weight parameters). The procedure here follows [40, 41]. (1) All the qubits are initialized to $|0\rangle$. (2) Specify the density matrix $S_1 = \rho^{in}$ of the input after the initialization step has been completed. (3) Tensor the density matrix with the inner product of the qubits in the first parameterized layer. At this point, our calculation takes the form: $S_2 = \rho^{in} \otimes |000\rangle \langle 000|$. (4) Apply the unitary matrices of the first parameterized layer to obtain $S_3 = IRy(\theta_2)Rz(\theta_1)(\rho^{in} \otimes |000\rangle \langle 000|)Rz^\dagger(\theta_1)Ry^\dagger(\theta_2)I$. Here, stands for the complex conjugate. (5) Tensor out the input layer by applying the trace operator to the expression in (4): $S_4 = tr_{in}\left(IRy(\theta_2)Rz(\theta_1)(\rho^{in} \otimes |000\rangle \langle 000|)Rz^\dagger(\theta_1)Ry^\dagger(\theta_2)I\right)$. This is possible because the unitary matrices of the subsequent layers will not affect the input layer. The trace operator sums up the elements on the main diagonal. (6) Tensor the expression in (5) with the inner product of the states of the second parameterized layer to obtain $S_5 = S_4 \otimes |000\rangle \langle 000|$. (7) Apply the unitary matrices of the second parameterized layer: $S_6 = Rx(\theta_4)IRy(\theta_3)S_5Ry^\dagger(\theta_3)IRx^\dagger(\theta_4)$. (8) Trace out the first parameterized layer by applying the trace operator: $S_7 = tr_1(S_6)$. This last state contains the values of the output.

Presently, the field of quantum neural networks is under active research with respect to both conceptual as well as implementation aspects of variational quantum circuits. Some considerations regarding these are for instance (1) how to efficiently embed classical data into quantum states, and (2) methodology for the case of so-called *barren plateaus*, which are large regions with gradient equaling 0 and thus not allowing for the calculation of gradient descent for the purpose of optimization.

5.2.6 Quantum Generative Adversarial Networks

Quantum generative adversarial network (QGAN) emerging in the domain of quantum machine learning as motivated by the success of standard Generative Adversarial Networks (GANs). In 2014, GAN was developed by Ian Goodfellow and improved by Alec Redford and various researchers [42].

GAN is a type of deep neural network for unsupervised machine learning in the domain of natural language processing, computer vision, and image analysis. The GAN model is classified into two sub-models: (i) Generator (G) and (ii) Discriminator (D).

- The goal of the generator is to generate new samples from the real domain such as text, audio, and images.
- The goal of the discriminator is used to classify and identify whether the generated are real or fake.

The generator receives a vector of random numerical as training input and converts it into an image using the Gaussian distribution. The discriminator model receives the generated images from the generator model to classify the predictions between 0 and 1, where 0 is a fake image and 1 is the original image.

Now, the difficulty with this technique is that a classical generator cannot handle quantum data effectively and must be converted back to classical data. Various researchers explored the studies to tackle and improve the performance of training issues in classical GAN using other algorithms such as Wasserstein GAN-Gradient

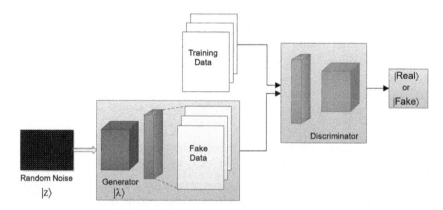

Figure 5.9 Architecture of Quantum generative adversarial network (QGAN).

Penalty (WGAN-GP), Self-Attention Generative Adversarial Networks (SAGAN), Deep Convolutional Generative Adversarial Networks (DCGAN), Boundary Equilibrium Generative Adversarial Networks (BEGAN), and Progressive Growing of Generative Adversarial Networks (PGGAN) [42]. Moreover, the choice of the metric between original and duplicate distributions will be a critical issue for the stability of the performance in the training process.

Quantum computing provides unique advantages over classical computing such as QGAN is the most important application due to its strong learning process in the training with limited parameters. QGAN is a data-driven quantum circuit machine learning algorithm that combines classical GAN and quantum computing [43, 44]. Lloyd et al. [54] and Weedbrook [26] firstly proposed the theory of QGAN based on the inherently probabilistic nature of the quantum system. Figure 5.9 shows an Architecture of QGAN. When quantum generators generate very high-dimensional measurement statistics, the QGAN can exhibit quantum advantages over the classical GAN. Therefore, QGAN can converge faster or require fewer physical resources. There are several applications of QGAN such as generating unique human faces, image-to-image translation, text-to-image translation, editing and repairing images, super resolution.

5.2.7 Quantum Autoencoders

The classical autoencoder is a special kind of unsupervised artificial neural network that learns to compress (encoding the data in a smaller dimension than the current one) data efficiently and also reconstruct the real data from the compressed one, as close as possible. This specific kind of model is categorized as "Unsupervised," because while training this neural network no label of the given data is incorporated, which is clarified in Figure 5.10.

The input layer of the neural network model consists of 784 neurons, which gets reduced to only 2 dimensions and further gets expanded to its original dimension, i.e., 784 neurons. The data compression part (here from 784 to 2) is termed as *Encoder* and the expansion part (here from 2 to 784) is termed as *Decoder*. The neural network model while being trained on a dataset, both the encoder and decoder part is trained together at the same time and that is the reason why it is an unsupervised model because both the input and output to the neural network is representing the same data, which can be directly compared as a cost value of the model and no other labeling of the data is required. The lowest dimensional layer at the middle is also called the *bottleneck* layer, which is quite self-explanatory. Importantly, the Encoder and Decoder part in either side of that bottleneck layer is usually designed as a mirror image of each other.

The overall concept of quantum autoencoder is similar to that of classical one. As shown in Figure 5.11, the initial state ρ^{in} represented by 4 qubits has been reduced to 2 qubits applying operator U, this is known as encoder [45, 46]. The exact inverse operation, i.e., applying $U^\dagger$ on those 2 qubits to revert back to the 2 qubits state is known as decoder.

In spite of this overall similarity, there is a fundamental difference between a classical and quantum neural network. Although in classical neural network, the number of neurons in each layer doesn't have to be exact same, but in quantum circuit the number of qubits cannot vary after each layer of operators. It is not possible to drop a qubit and initialize a new one in between the execution (at least in current systems) [47]. Therefore, the measurement and initialization of 2 qubits in the bottleneck part of the autoencoder, as shown in Figure 5.11.

Let's say the initial state ρ^{in} is represented by the total $(n+k)$ number of qubits. The unitary operator U, which consists of multiple trainable parameterized gates (RX, RY, RZ, etc.) and entangling gates (CNOT, CZ, etc.) has been applied to the $(n+k)$ qubits. The choice of the gate design of the operator U is quite dataset specific. After this operation, from $(n+k)$ qubits, first n qubits will be dropped (practically it is not possible, discussed in the next section) keeping the k qubits as encoded state.

Then in place of the dropped n number of qubits, new qubits are initialized and the inverse of operator U, which is represented as $U^\dagger$ has been applied. Finally, the final state ρ^{out}, which is of the same dimension as ρ^{in}, are compared with each other. For an ideal autoencoder, it will be $\rho^{in} = \rho^{out}$. Therefore the loss function to train the autoencoder can be theoretically described as $<\rho^{in} \mid \rho^{out} \mid \rho^{in}>$, which should

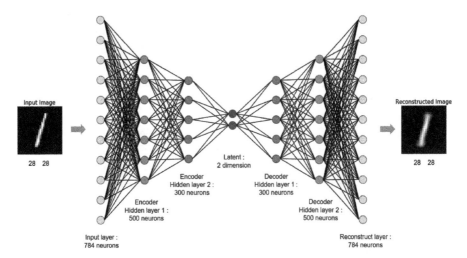

Figure 5.10 The schematic representation of classical autoencoder model.

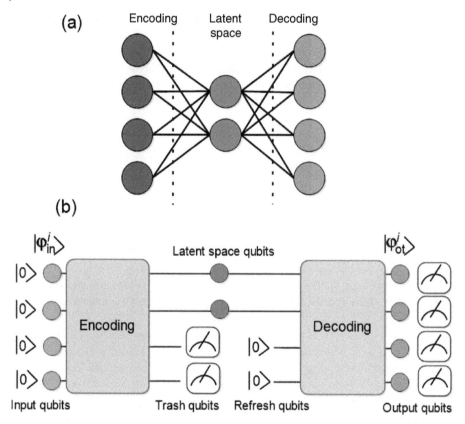

Figure 5.11 The schematic representation of classical and quantum autoencoder model with four number of qubits.

approach to the value 1, with the circuit being trained. The overall cost value of a circuit with parameters p, calculated on dataset with i number of samples can be written as $C_1(p) = \Sigma_i\, p_i\, F(\rho^{in}, \rho^{out})$, where $F(\rho^{in}, \rho^{out}) = <\rho^{in} \mid \rho^{out} \mid \rho^{in}>$. Now the variational circuit being defined and cost being calculated, the quantum autoencoder model will be trained for a specific number of epochs, until the cost of circuit converges to a stable point [27]. After training, the encoder and decoder part are being used separately as done in classical. $(n+s)$ qubits data will be reduced to only k qubits using the encoder and vice versa with the decoder. The experimental demonstration of quantum autoencoder is performed and qutrits are compressed using machine learning models [48].

Now coming to the issue of new qubits not being possible to be initialized in between the execution of a circuit as stated previously, there is a trick that has been applied to overcome that, but that demands extra n number of qubits. Here we have kept extra n number of qubits shown by state la> in Figure 5.12 that are initialized as zero state. Next $(n+k)$ qubits represents state ψ_i where n number of qubits are denoted by B and k number of qubits by X in Figure 5.11. After applying the unitary operator U on the state ψ_i, Y and Y' has been swapped with each other using swap gates. Then, finally the operator $U^\dagger$ is applied on the $(n+k)$ qubits to revive ρ^{out}. After swapping Y

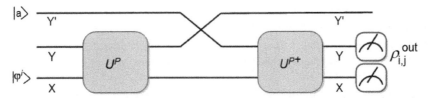

Figure 5.12 The working of bottleneck layer in quantum autoencoder to compare the qubits with the measured qubits.

and Y', the Y' qubits which are not further used with any operator (for this reason it's also known as trash qubit), are comparable with the measured qubits at the bottleneck layer in Figure 5.11. Similarly after swap, the Y qubits are comparable with initialized zero states in bottleneck layer of Figure 5.11 because Y qubits after swap basically represents the reference state |a>, which was initialized to zero states for all *n* qubits.

5.3 Conclusion

The chapter provides an overview of recent progresses and different aspects of an emerging area of quantum machine learning. The small and larger special-purpose quantum computers, simulators, annealers have shown a great capability to use in data analytics and machine learning. The quantum machine learning algorithms are based on HHL or quantum amplitude approach, Grover's searching algorithm or quantum annealing methods. The integration of machine learning, gradient-based methods and Noisy Intermediate-Scale Quantum (NISQ) devices is still a young area and potentially has a lot more to offer. In the future, this emerging field will open up new avenues of research in solving classical and quantum optimization problems and framework design as quantum technologies become more mature.

Bibliography

1 Anderson, J. R. (1990). *Machine Learning: An Artificial Intelligence Approach*, Vol. 3. Morgan Kaufmann.

2 Mitchell, R., Michalski, J., and Carbonell, T. (2013). *An Artificial Intelligence Approach*. Berlin: Springer.

3 Marsland, S. (2015). *Machine Learning: An Algorithmic Perspective*. CRC press.

4 Ayodele, T. O. (2010). Types of machine learning algorithms. *New Advances in Machine Learning* 3: 19–48.

5 Jordan, M. I. and Mitchell, T. M. (2015). Machine learning: trends, perspectives, and prospects. *Science* 349 (6245): 255–260.

6 Gruska, J. (1999). *Quantum Computing*, Vol. 2005. London: McGraw-Hill.

7 McMahon, D. (2007). *Quantum Computing Explained*. John Wiley & Sons.

8 Nielsen, M. A. and Chuang, I. (2002). *Quantum Computation and Quantum Information, Cambridge University Press.*

9 Grover, L. K. (1996, July). A fast quantum mechanical algorithm for database search. *Proceedings of the Twenty-Eighth Annual ACM symposium on Theory of Computing*, 212–219.

10 Preskill, J. (2018). Quantum computing in the NISQ era and beyond. *Quantum* 2: 79.

11 Cárdenas-López, F. A., Sanz, M., Retamal, J. C., and Solano, E. (2019). Enhanced quantum synchronization via quantum machine learning. *Advanced Quantum Technologies* 2 (7-8): 1800076.

12 Biamonte, J., Wittek, P., Pancotti, N., Rebentrost, P., Wiebe, N., and Lloyd, S. (2017). Quantum machine learning. *Nature* 549 (7671): 195–202.

13 Bhatia, A. S., Saggi, M. K., Kumar, A., and Jain, S. (2019). Matrix product state–based quantum classifier. *Neural Computation* 31 (7): 1499–1517.

14 Huggins, W., Patil, P., Mitchell, B., Whaley, K. B., and Stoudenmire, E. M. (2019). Towards quantum machine learning with tensor networks. *Quantum Science and Technology* 4 (2): 024001.

15 Orús, R. (2019). Tensor networks for complex quantum systems. *Nature Reviews Physics* 1 (9): 538–550.

16 Sen, P., Bhatia, A. S., Bhangu, K. S., and Elbeltagi, A. (2022). Variational quantum classifiers through the lens of the Hessian. *Plos one* 17 (1): e0262346.

17 Stoudenmire, E. M. and Schwab, D. J. (2016). Supervised learning with quantum-inspired tensor networks. arXiv preprint arXiv:1605.05775.

18 Rebentrost, P., Mohseni, M., and Lloyd, S. (2014). Quantum support vector machine for big data classification. *Physical Review Letters* 113 (13): 130503.

19 Wiebe, N., Kapoor, A., and Svore, K. (2014). Quantum algorithms for nearest-neighbor methods for supervised and unsupervised learning. arXiv preprint arXiv:1401.2142.

20 Aïmeur, E., Brassard, G., and Gambs, S. (2013). Quantum speed-up for unsupervised learning. *Machine Learning* 90 (2): 261–287.

21 Lloyd, S., Mohseni, M., and Rebentrost, P. (2013). Quantum algorithms for supervised and unsupervised machine learning. arXiv preprint arXiv:1307.0411.

22 Li, R. Y., Di Felice, R., Rohs, R., and Lidar, D. A. (2018). Quantum annealing versus classical machine learning applied to a simplified computational biology problem. *npj Quantum Information* 4 (1): 1–10.

23 Wiebe, N., Kapoor, A., Granade, C., and Svore, K. M. (2015). Quantum inspired training for Boltzmann machines. arXiv preprint arXiv:1507.02642.

24 Anguita, D., Ridella, S., Rivieccio, F., and Zunino, R. (2003). Quantum optimization for training support vector machines. *Neural Networks* 16 (5-6): 763–770.

25 Amin, M. H., Andriyash, E., Rolfe, J., Kulchytskyy, B., and Melko, R. (2018). Quantum Boltzmann machine. *Physical Review X* 8 (2): 021050.

26 Lloyd, S. and Weedbrook, C. (2018). Quantum generative adversarial learning. *Physical Review Letters* 121 (4): 040502.

27 Khoshaman, A., Vinci, W., Denis, B., Andriyash, E., Sadeghi, H., and Amin, M. H. (2018). Quantum variational autoencoder. *Quantum Science and Technology* 4 (1): 014001.

28 Dang, Y., Jiang, N., Hu, H., Ji, Z., and Zhang, W. (2018). Image classification based on quantum K-Nearest-Neighbor algorithm. *Quantum Information Processing* 17 (9): 239.

29 Ruan, Y., Xue, X., Liu, H., Tan, J., and Li, X. (2017). Quantum algorithm for k-nearest neighbors classification based on the metric of hamming distance. *International Journal of Theoretical Physics* 56 (11): 3496–3507.

30 Wang, Y., Wang, R., Li, D., Adu-Gyamfi, D., Tian, K., and Zhu, Y. (2019). Improved handwritten digit recognition using quantum K-nearest neighbor algorithm. *International Journal of Theoretical Physics* 58 (7): 2331–2340.

31 Abdi, H. and Williams, L. J. (2010). Principal component analysis. *Wiley Interdisciplinary Reviews: Computational Statistics* 2 (4): 433–459.

32 Wold, S., Esbensen, K., and Geladi, P. (1987). Principal component analysis. *Chemometrics and Intelligent Laboratory Systems* 2 (1-3): 37–52.

33 Lloyd, S., Mohseni, M., and Rebentrost, P. (2014). Quantum principal component analysis. *Nature Physics* 10 (9): 631–633.

34 LaRose, R., Tikku, A., O'Neel-Judy, É., Cincio, L., and Coles, P. J. (2019). Variational quantum state diagonalization. *npj Quantum Information* 5 (1): 1–10.

35 Ostaszewski, M., Sadowski, P., and Gawron, P. (2015). Quantum image classification using principal component analysis. arXiv preprint arXiv:1504.00580.

36 Schuld, M., Sinayskiy, I., and Petruccione, F. (2015). An introduction to quantum machine learning. *Contemporary Physics* 56 (2): 172–185.

37 Kulchytskyy, B., Andriyash, E., Amin, M., and Melko, R. (2016). Quantum boltzmann machine. APS Meeting Abstracts.

38 Abbas, A., Sutter, D., Zoufal, C., Lucchi, A., Figalli, A., and Woerner, S. (2020). The power of quantum neural networks. arXiv preprint arXiv:2011.00027.

39 Mari, A., Bromley, T.R., and Killoran, N. (2021). Estimating the gradient and higher-order derivatives on quantum hardware. *Physical Review A* 103 (1): 012405.

40 Beer, K., Bondarenko, D., Farrelly, T., Osborne, T.J., Salzmann, R., Scheiermann, D., and Wolf, R. (2020). Training deep quantum neural networks. *Nature Communications* 11 (1): 1–6.

41 Farhi, E. and Neven, H. (2018). Classification with quantum neural networks on near term processors. arXiv preprint arXiv:1802.06002.

42 Goodfellow, I., Pouget-Abadie, J., Mirza, M., Xu, B., Warde-Farley, D., Ozair, S., ... Bengio, Y. (2014). Generative adversarial nets. *Advances in Neural Information Processing Systems* 27.

43 Benedetti, M., Garcia-Pintos, D., Perdomo, O., Leyton-Ortega, V., Nam, Y., and Perdomo-Ortiz, A. (2019). A generative modeling approach for benchmarking and training shallow quantum circuits. *npj Quantum Information* 5 (1): 1–9.

44 Chakrabarti, S., Huang, Y., Tongyang, L., Feizi, S., and Wu, X. (2019). Quantum Wasserstein GANs. *33rd Conference on Neural Information Processing Systems (NeurIPS 2019)*.

45 Bravo-Prieto, C. (2020). Quantum autoencoders with enhanced data encoding. arXiv preprint arXiv:2010.06599.

46 Romero, J., Olson, J. P., and Aspuru-Guzik, A. (2017). Quantum autoencoders for efficient compression of quantum data. *Quantum Science and Technology* 2 (4): 045001.

47 Bondarenko, D. and Feldmann, P. (2020). Quantum autoencoders to denoise quantum data. *Physical Review Letters* 124 (13): 130502.

48 Pepper, A., Tischler, N., and Pryde, G.J. (2019). Experimental realization of a quantum autoencoder: the compression of qutrits via machine learning. *Physical Review Letters* 122 (6): 060501.

49 Zhang, P., Li, S., and Zhou, Y. (2015). An algorithm of quantum restricted boltzmann machine network based on quantum gates and its application. *Shock and Vibration* 2015.

50 Lloyd, S. (2010). Quantum algorithm for solving linear systems of equations. *In APS March Meeting Abstracts* 2010: D4–002.

51 Dong, D., Chen, C., Li, H., and Tarn, T. J. (2008). Quantum reinforcement learning. *IEEE Transactions on Systems, Man, and Cybernetics, Part B (Cybernetics)* 38 (5): 1207–1220.

52 Aimeur, E., Brassard, G., and Gambs, S. (2007). Quantum clustering algorithms. In *Proceedings of the 24th international conference on machine learning*, (pp. 1–8), Corvalis Oregon, USA.

53 Wittek, P. (2014). *Quantum machine learning: what quantum computing means to data mining*, First Edition, Academic Press.

54 Lloyd, S., and Weedbrook, C. (2018). Quantum generative adversarial learning. *Physical review letters*, 121 (4): 040502.

Part 3

Computational Intelligence and Its Applications

6

Computational Intelligence Paradigms in Radiological Image Processing—Recent Trends and Challenges

Anil B. Gavade[1], Rajendra B. Nerli[2], Ashwin Patil[3], Shridhar Ghagane[4] and Venkata Siva Prasad Bhagavatula[5]

[1]Department of E&C, KLS Gogte Institute of Technology, Belagavi, Karnataka, India
[2,3,4]Department of Urology & Radiology, JN Medical College, KLE Academy of Higher Education and Research (Deemed-to-be-University), Belagavi, Karnataka, India
[5]Medtronic, Hyderabad, India

6.1 Introduction

A current boom in the modeling intelligence in algorithm to solve complex applications, this intelligence could be achieved through natural and biological intelligence, resulted a technology known as intelligent systems, these algorithms use soft computing tools. AI aim to make the machines and computers smarter, that make a computer to mimic like human brain in specific applications. AI algorithms are a blend of many research areas, such as biology, sociology, philosophy, and computer science. The purpose of AI is not to substitute human beings, instead offer us a more prevailing tool to support in our work, provide more computing ability, permitting them to exhibit more intelligent behavior.

6.2 Computational Intelligence

CI is a fragment of AI, which deals with study of adaptive mechanisms to enable intelligent behaviour in complex changing environment Figure 6.1. Shows all paradigm. Three main columns of CI are Fuzzy Systems, Neural Networks, and Evolutionary Computation, individual components have certain weakness and these could be improved by combining them tougher and they are referred as hybrid CI. Computers learn specific tasks from diverse forms of data or experimental observation, the ability to make computers learn and adapt is usually referred to as CI. It is considered to have the ability of computational adaptation, high computational speed, and fault tolerance. Computational adaptation means the ability of a system to adapt the changes happens in its input and output instances.

6.2.1 Difference between AI and CI

AI deals with study of intelligent behavior exhibited by machines, mimicking natural intelligence similar to humans, AI aims to develop an intelligent machine which can think, act, and take decision similar to human. CI is the study of adapting and building intelligent behaviors based on a changing complex environment. CI is to recognize the computational model which make intelligent behavior of artificial and

Emerging Computing Paradigms: Principles, Advances and Applications, First Edition.
Edited by Umang Singh, San Murugesan and Ashish Seth.
© 2022 John Wiley & Sons Ltd. Published 2022 by John Wiley & Sons Ltd.

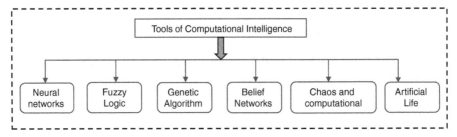

Figure 6.1 Computation intelligence paradigm.

natural system in complex environment. AI and CI and have nearly similar type of goals but they are moderately different from each other.

6.2.2 Tools of Computational Intelligence

CI is the theory design, application, and development of a biologically motivated computational model. Traditionally it has three pillars, they are neural networks, fuzzy system, and evolutionary computational. It encloses computing models like artificial life, culture learning, social reasoning, and artificial hormone network. CI plays a major role in developing successful intelligent systems that includes games and a cognitive developmental system. In reality, some of the greatest successful AI systems depend on CI.

6.3 Radiological Image Processing Introduction

Information is knowledge and it can be represented in different forms, digital image or image is one of them, image is worth 1,000 words. Most commonly, humans depend on images perceived by our eyes more than any other sensory stimulus. Human eyes capture the image, the brain extracts information and interprets the objects. Today, most of the computer vision and machine learning applications are working on the similar lines. The drastic improvement and proliferation of radiological imaging has changed the medicine, allowed physicians and scientists to gather information by looking noninvasively into the human body. Medical imagery role has extended beyond visualization and inspection of anatomic structures, acting as new tool for surgical planning and simulation, intraoperative navigation, disease progress tracking, and radiotherapy planning etc.

Medical diagnostics, today, extensively depend on direct digital imaging techniques, almost all radiological modalities are available in the digital formats. Complexity of information differs from one modality to other that ranges from X-ray to MRI or an ultrasound image of an organ. Radiological application started with analog imaging modalities and today all are in digital format due to improvement in sensor and computation technology, almost all radiological application is in digital today. Medical images efficiently processed, objectively assessed and accessible at several places at same time through protocols and communication networks, such as Digital Imaging and Communications in Medicine (DICOM) and Picture Archiving and Communication Systems (PACS).

Digital image processing is an area of science and mathematics that manipulate the information present in the image, after manipulation the results could be input to several applications. We find digital image processing applications in almost several areas of engineering and science that ranges from space exploration, robotics to medical applications. Digital image is a two-dimensional function that is represented in a matrix, as rows and columns, the smallest entity is referred as pixel/pel. The stages in digital image processing involves acquisition of image using sensors such as a charge-coupled device (CCD), store and process using digital computer and finally display or print. Processing of radiological digital image involves image enhancement, image restoration, image analysis, image segmentation, image compression, image synthesis, and image quantification. Digital image is represented in different forms as black and white, grey scale, color and compressed images, image resolution, and type of image are directly correlated with data dimension. A medical image is commonly blurry and noisy due to acquisition stages, the degradation of the image is due to poor contrast or illumination and noise. Biomedical image analysis involves several stages and these stages are commonly requirements for subsequent stages, the final stage involves storage and decision making on capture image by medical practitioner or by machine to assist radiologist. These stages involved are image acquisition, image enhancement and restoration, image segmentation, image classification and quantification, to perform these algorithms more efficiently and precisely they need to be intelligent and fast in computation, CI is the best tool for these algorithms.

Radiological imaging is a vastly interdisciplinary field, which is combination of physics, medicine, computer sciences, and engineering. Primarily, radiological/biomedical imaging analysis is relevance application of digital image processing to medical or biological problems. However, in radiological image application a number of other fields play a vital role, like physiology, anatomy, and physics of the imaging modality and instrumentation, etc. The diagnosis or interference in medical application delivers the basis and motivation for biomedical image analysis. The choice of an imagery modality and of possible image processing stages depends on various medical factors, such as the type of tissue to be imaged or the suspected disease. Radiological imagery applications consist of four different stages, generally each stage is connected to the future subsequent stages, but at any required stage the algorithms allow human to human intervention to make decision or the results could be recorded. Imagery application have these minimum stages as image capturing (acquisition), image enhancement and restoration, image segmentations and classification, and finally image quantification shown in Figure 6.2.

6.3.1 Image Acquisition

Image acquisition is the first step to form the 2-dimension object digitally, such as a suspicious tissue in a patient, spatial resolution is significant in biomedical imageries. Digital image is mapping of one or several tissue properties on the discrete quadrangular grid, these grids are pixels voxel (volume element) in 3-dimensional images, these discrete values are stored in memory as integer values. Each pixels or voxels have physical meaning, example endoscopic and photography image values that are relative to light intensities. Computed Tomography (CT) carries image values

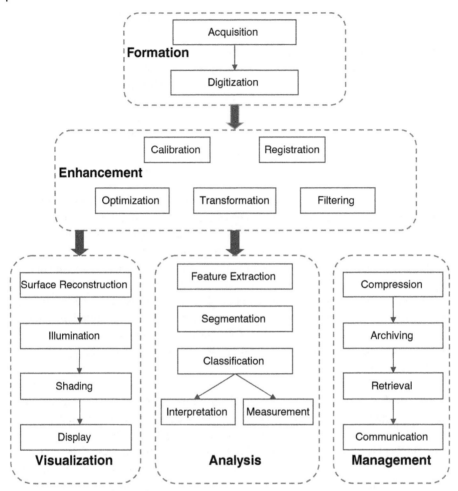

Figure 6.2 Digital radiological image processing.

that are relative to local X-ray absorption. In MRI, the image values can represent a variety of tissue properties, liable on the acquisition sequence, essentially times proton density or local echo decay. The aim of the image acquisition step is to acquire contrast of the tissues, to analyze. The human eye is enormously superior at recognizing and classifying meaningful contrast, even in condition with poor signal-to-noise (S/N) ratio. Human vision permits immediate recognition and identification of spatial associations and makes it conceivable to notice subtle differences in density and to filter the feature from the noise. Experienced radiologist will have no trouble in identifying normal and abnormal tissues from the radiological digital images, but for computer it is a challenging task to do, next steps of image processing steps involve into the role and they are more significant for automation.

Image Enhancement: In computer vison and machine learning application, resolution of image is substantial for two purposes, first it improves the perception visibility features for more accurate and precise diagnosis by radiologist. Secondly, subsequent stage performs in best possible ways like segmentation, identification, classification, and quantification image data. Pixel value remapping, filtering (spatial and frequency), and few restoration methods are most commonly used enhancement operators. Histogram equalization and histogram stretching are two linear and non-linear enhancement techniques. Filters attenuate/amplify relevant characteristics of pixels in an image, and filters make use of pixel vicinity. Filters work on pixels known as spatial domain filters and those which uses transforms such as discrete cosine transform, fourier transform and wavelet transform etc., which defines image data in terms of periodic components they all come in frequency domain filters. Filter play a vital role to improve smoothing an image, sharpening edges of objects in an image, and suppress periodic artifacts, to eliminate in an heterogenous background intensity distribution.

Image restoration and enhancement, they work on the similar line to improve the degradation in image. Image degradation occurs to due to misfocus of lenses, noise and due to motion blur of camera, degradation of image occurs at the acquisition process itself. To reverse the degradation of image we require filters and these filters work on reverse degradation process, such that errors (mean-squared error) is reduced between the restored image and idealized unknown image. Application of microscopy imaging inhomogeneous illumination play a significant role; this illumination may vary over time and this leads to introduction motion artifacts. While filters employed to overcome blur effect, they require well stable design measures, otherwise local contrast enhancement led to increase noise component and therefore decrease in the signal-to-noise ratio. Equally noise reducing filters negatively affect details of texture and edges, while these filters enhance the signal-to-noise ratio, details image may be lost and get blurred. Design of enhancement filter depend on the further steps of image processing stage. Enhanced image is always preferred may be for machine perception or human observation, human eyes distinguish particular objects from image even there is significant noise exist, but machines can't do the same. Improving resolution of image is possible by several techniques. Super Resolution (SR) image and Image Fusion (IF) are commonly employed methods in number of applications.

6.3.2 Super Resolution Image Reconstruction

Super Resolution (SR)is a process in which perceptual quality of image is improved by replicating neighbouring pixels, zooming pixels, or by adding multiple frames of same image to be reconstructed. Machine learning approaches contribute a large number algorithm, which is shown in Figure 6.3. Currently, the efficiency and accuracy of SR techniques based on machine learning reached a stage where high-end computational resources high resolution image reconstruction is conceivable. The choice of the finest suited algorithm must consider number of aspects, such as medical imaging, robotic supervision of objects or satellite imagery applications, etc. It is always essential to apply external learning, in terms of memory requirement and computational complexity allowed i.e., accuracy trade off and processing time. Over all the selection of SR algorithm to each problem need a careful attention of the limitations presented by the application situation. Biomedical image enhancement is a process to remove and reduce the artifacts originated due to improper illumination ambience, several researchers addressed SR in transform domain, it is observed few

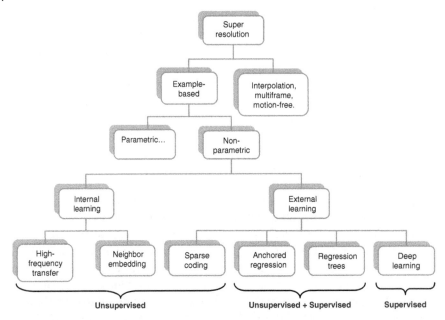

Figure 6.3   Machine learning methods in super resolution image enhancement.

pixels are lost in the process of transformation from one domain to other. Edges of objects i.e., diseased tissues involve a significant role in image analysis, identification and vision processing. As pixels are lost due transform domain mapping edges of tissues, boundaries, and textures are degraded that leads to inappropriate diagnosis. Image enhancement algorithms are always application specific; these has to be addressed the specific radiologist sensitive to contrast ratio and their preference has to be evaluated.

6.3.3   Image Fusion

Image fusion is a process of combining multiplex images, from multiple sensors and combined to form single image, always this image is having more information than the individual one, and consist all relevant information. The aim of image fusion is to construct better quality, more suitable for machine and human perception. Image fusion is carried out as pixel fusion, feature fusion, and decision fusion. Image fusion has tremendous demand in the area of radiological diagnostics followed with appropriate treatment. The term multi-image fusion refers to combining multiple images of a patient from the same modalities or images taken from different modalities like MRI and CT image, etc., different image fusion techniques are available today shown in Figure 6.4.

6.3.4   Image Restoration

Image restoration is a mathematical operation on corrupted digital image and estimate clear original image, the degradation of image may be due to object camera misfocus, noise, random atmospheric turbulence, and motion blur. Image enhancement and

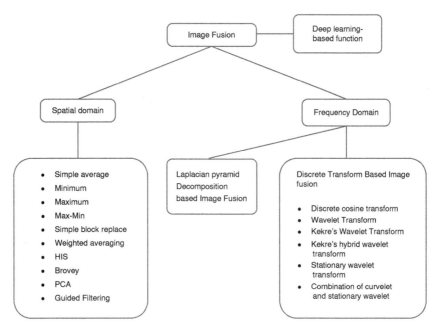

Figure 6.4　Image fusion techniques.

image restoration are not same, this process improves features of the image which are informative in further stages. Thus, image restoration concentrates to noise function and modeling blurring, then applying inverse model to de-blurred and de-noise the image. The objective of image restoration is to develop restoration algorithm to filter and eliminate the degradation from input image in doing this soft computing and computation intelligence play vital role.

Digital image restoration deal with method used to suppress a known degradation to recover an original image. It's upcoming field of an image processing. The objective of image restoration is to restore distorted/degraded image to its original quality and content. Degradation is introduced by an image acquisition device due to nonlinearity of sensors, detects in optical lenses, blur due to misfocus of camera, relative object camera motion, atmospheric turbulence, etc.

It tries to minimize some parameters of degradation and reconstructing an image that has been degraded based on known degradation (prior knowledge) models and mathematical or probabilistic models. Usually, iterative restoration techniques attempt to modeling degradation and then applying the inverse process to recover the original image.

There are two subprocesses:

i)　By adding noise and blur to an image degrading the quality of an image.
ii)　Recovering the original image

In restoration application deblurring is very important because visually blurring is annoying. The different kind of filters and additive noise are used for blurring an image. Quality of image is degrading by adding Gaussian and salt pepper noise, as shown in Figure 6.5.

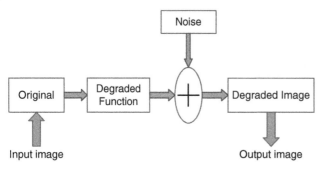

Figure 6.5 Degradation model for blurring the image.

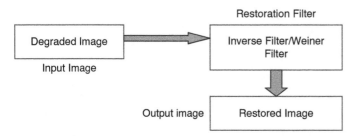

Figure 6.6 Restoration model.

6.3.5 Restoration Model

This process estimates from degraded version using filter restoration blur and noise image factor is removed in order to obtained original image, as shown in Figure 6.6.

Ma. Guadalupe Sanchez et al. [1]. In the recent decade, several optimization methods are proposed depending on the type of noise. This paper explained the algorithm to remove Gaussian, speckle, and impulsive noise. NDF, PGFM, and PGFND methods were used for filtration purposes and compared obtained quality result in each case. If the Gaussian and speckle noise is present NDF method perform good to reduce the noise. If the impulsive noise (fixed) is appeared in an image then the best technique is PGFM and to deal with the combination of various noises PGFND techniques is able to reduce the noise more effectively.

A. Lakshmi and Subrata Rakshit [2] described object evaluation method were analyzed by comparing proposed and distortion measures with different restoration algorithm for the estimation of undistorted image which automates the process of restoration in real time without demanding any kind of knowledge about original image and it's derived without any assumption of image statistics and noise. The given measures have noise assessing terms and data fidelity term thus analyzing denoising as well as deblurring nature of image restoration method.

A.M. Raid [3] present image restoration based on morphological operation. There are two main morphological operations i.e, dilation and e,rosion. In dilation operation the object is expanded, thus small holes are filled and it connects to disjoint objects. The proposed methodology mainly focuses on two basic morphological algorithms (region filling and boundary extraction) and four morphological operations (opening and closing, dilation and erosion). It's implemented using the MATLAB program with user interface which changing the SE parameters such as its type or size are simple but if the objects are near with the distance then it will be stuck together, thus it's need to solve this program by searching objects.

6.3.6 Image Analysis

Image analysis is an extraction of meaningful information from input image, with these information algorithms will able to identify the objects, the more feature we provide the better classification and identification we can achieve. In achieving high accuracies, the algorithm needs to be more intelligent and faster to respond, these can be achieved using deep learning, neural network, capsule network, and computation intelligence.

6.3.7 Image Segmentation

Image segmentation is one of the most significant steps in digital image processing, it is to separate foreground and background, which are treated as different objects in an image. The goal of segmentation is to classify each of the pixel as one of the classes and extracting region of interest (ROI). To achieve image segmentation effectively, the object must be different from one other, such as boundary, image intensity, texture, and shape etc. The aim of this stage may be either an outline or a mask i.e., outline may be a set of curves or parametric cure, like polygonal approximation of an object outline (shape), mask is to assign pixel value 1 to objects and treat back ground pixels as 0 or vice versa. Image segmentation is one of the most complicated tasks, we have several segmentation methods and they are more application specific. Segmentation process should stop when the region of interest has been isolated. An outline of the most widespread segmentation techniques as follows, shown in the Figure 6.7.

Edge-based Segmentation: Edge detection is a mathematical operation is an image processing, it detects boundaries of objects present in an image. It works with detecting sharp changes in intensity of the pixel that typically forms boarder between different object. Basic idea of edge detection is look for neighborhood pixel with strong sign of change. These pixels are easily detected by computer on basis of intensity differences. Mainly it's process of finding meaningful transaction in an image. Feature extraction, image morphology, and pattern recognition can be achieved by edge detection. It extracts the features such as corners, lines, and curves of an image. Therefore, it is very easy to recognize segmentation boundaries and objects. Edge

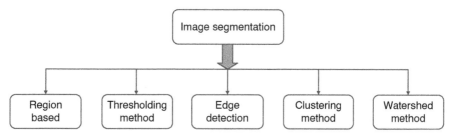

Figure 6.7 Different image segmentation.

detection steps consist smoothing, enhancement, detection and localization of an image. Commonly used edge detection types are step edge, ramp edge, ridge edge, and roof edge.

Edge Detection Approaches: Spatial domain and frequency domain are two classes of edge detection. Spatial domain includes operator-based approaches categorize into first order and second order method. First order methods are Prewitt, Sobel, and Robert. Cany and Laplacian are second order methods. Fourier transform method is used to convert the image into frequency domain. Using the low frequencies, the details were extracted from the image. High frequencies were used to obtain image edges, but frequencies having certain limitations.

Gradient-based Edge Detection: It is a first order derivative which compute gradient magnitude horizontally and vertically. Implementation of gradient-based edge detection method is very simple and capable of detecting edges and their directions. But edges are not located accurately because it is sensitive to noise.

Sobel Edge Detection Operator: This operator extracts all the edges without worrying about the directions. It computes gradient approximation of image intensity. It uses the 3×3 kernel convolved with input image to compute vertical and horizontal approximation, respectively. It provides smoothing effect and time efficient computation. But it has certain limitations, its highly sensitive to noise, not very accurate because it does not give appropriate result on thick and rough edges.

Prewitt Operator: Prewitt edge detection operator detects vertical and horizontal edges of an image. It uses kernels or masks. It is a best operator to detect magnitude and orientation of an image.

Robert Edge Detection Operator: It is used to compute the sum of square of difference between diagonally adjacent pixels in an image through discrete differentiation. In this operator orientation and detection of edges are very easy. It preserves diagonal direction point but it's very sensitive to noise therefore not accurate method for edge detection.

Laplacian of Gaussian (LOG): It is derivative operator that uses the Laplacian take as a second derivative of an image. It is used to find sharp edges of all directions having a fixed characteristic and easily detects the edges.

Canny Operator: It is a Gaussian-based operator. Canny operator is most commonly used because it can extract features of an image without altering the features as well as it localizes the edge points and less sensitive to noise.

Thresholding Based Segmentation: Threshold is the simplest and most powerful technique in a image segmentation. In this technique depending on intensity value the pixels are partitioned. Based on threshold value from the grey scale image it produces the binary image. It has a advantages such as fast processing speed, smaller storage space, and ease in manipulation compared with grey level image. Therefore, thresholding technique is commonly used.

Clustering: A collection of an arrangement of similar items is called clustering. The clustering method groups together a similar pattern and can produce a very good segmentation result. In clustering it is frequently important to change the information during the preprocessing and demonstrate the parameters until the outcome accomplishes the desire properties. There are different clustering methods, K-means, fuzzy C-means, mixture of Gaussians, and ANN clustering.

- **K-means** clustering method uses unsupervised algorithm. The result is well separated. K-means is fast and robust but it has noisy data and a nonlinear dataset.
- **Fuzzy C-means** uses an assigning membership algorithm with a cluster center. But the result is overlapped and comparatively its better than K-means.
- **Mixture of Gaussian** algorithm is based on a priori "n" Gaussian. Where all data taken is the minimum and maximum from Gaussian centers. This method is best for real-world data but its complex in nature.
- **ANN clustering** is based on priori data. Whose result were well separated. An ANN is mainly working on noisy image but it has slow convergence rate.

6.3.8   Region-Based Segmentation

The region-based segmentation is used to classify a particular image into number of regions or classes. Therefore, we need to estimate and classify each pixel in the image. Methods of region-based segmentation are region growing, texture-based segmentation and edge-based snakes. Region growing is general technique of image segmentation, where image characteristics are used to group neighboring pixel together to form regions. Region based techniques look for consistency within a subregion based on a property like intensity, color, texture etc. Region based segmentation starts in the middle of an object and then grows slowly towards till it meets the object boundary.

6.3.9   Watershed Segmentation

Watershed segmentation is morphology segmentation, that use watershed transform, that belongs to a region-based segmentation technique. Watershed transform can get one-pixel wide connected closed, accurate positioning of the edge. This algorithm is automatic and doesn't require any parameters to determine the termination conditions.

Namata Mittal et al. [4] In this study, efficient edge detection approaches analyzed for image analysis. Proposed method is tested on normal and medical images and compared with all traditional edge detection algorithms. Developed method is capable to obtain better entropy value and edge continuity along with less noise proportion. It is not effective for blurry images and time consumption need to improve. The problem encountered in traditional edge detection technique i.e., connectivity and edge thickness can be solved by B-Edge, that uses multiple threshold approaches and for effective edge detection and better connectivity the proposed methodology uses the triple intensity threshold value. Finally, it was concluded that B-Edge obtained a better outcome than canny. Developed method is able to perform good connectivity with improved edge width uniformity and produces acceptable entropy value.

Yousif A Hamad et al. [5] notes a low contrast medical image edge detection based fuzzy C-mean clustering have been developed. A canny edge detection algorithm performs well among all edge detection techniques. A FCM clustering segmented the image. Algorithm and software will develop in order to provide image analysis to its primary stage. To solve urgent diagnosis problem more analysis and processing will be done for real-time clinical CT and MRI imagery.

Tessy Badriyah et al. [6] notes that stroke classification is conducted in this study. CT brain imagery consist of more noise. Mainly in thresholding the gray scale image is converted into binary image to segment the affected tissues from CT image. Here global threshold and Otsu threshold is used for classification of stroke i.e., ischemic stroke classes, no stroke (normal) and hemorrhagic stroke. The proposed methodology is experimented by analyzing three filters i.e., Gaussian filter, bilateral filter, and median filter to remove noise. Quality of image is improved using peak signal-to-noise ratio i.e., 69% and mean–square error i.e., 0.008% with bilateral filter. Otsu thresholding is used for stroke object segmentation by specifying lower threshold parameter ≤ 170.

Alexander Zotin et al. [7] describes the proposed methods to detect the brain tumor from patients MRI scan image. In the first step, noise removal functions like median filter were used to improve features of medical images for reliability and enhancing balance contrast enhancement techniques (BCET). This image is segmented by a fuzzy C-means method and canny edge detector is applied to construct the edge map of brain tumor. In this paper they have compared sensitivity and accuracy parameters of different detection methods with proposed method. After comparing they have combined cany and fuzzy C-means together for better accuracy and sensitivity than the other single methods.

Cui, Xuemei et al. [8] has proposed algorithm based on an improved watershed transform method. Watershed transform has good response to the weak edges, but it is unable to obtain meaningful segmentation result directly. Therefore, they have made some improvements. Here they have briefly explained about image segmentation, watershed algorithm, and marker extraction. Watershed transform is the morphology segmentation method, mainly used for the study of the forms shape or structure of things. This method can suppress the noise and fine texture very accurately by avoiding over segmentation.

Abubakar et al. [9] has described about two categories of image segmentation. Here they have explained about sobel, canny and Robert cross field. Image segmentation is a segmentation of an image that is used in separating the object of an image from its background. From the experiment it was observed that canny edge detector better edge detection maps. And other image thresholding success fully separated the frequency from background.

Dubey et al. [10] gave a brief review about the image segmentation using different clustering methods. Clustering is the collection of an arrangement of articles such that items in a similar gathering called cluster. Here they have described about different clustering techniques that are K-means, fuzzy C-means, ANN clustering, and a mixture of Gaussians. By doing comparative study among these techniques by taking some important parameters such as data center, algorithm used, advantages and disadvantages, and final best result. After comparing they have concluded that fuzzy C-means is better than K-means. The mixture of Gaussian is used in real-world data.

Xu Gongwen et al. [11] describes wavelet transform base medical image segmentation as a broad term that encompasses a large range of applications. In this paper, analyzed frequency and time domain tool with some good features. It reduces noise and pointed the edges more precisely. The derived model solved all the problems that occur during traditional and classic algorithm. More analysis can be done that will enable us to improve new algorithm for segmentation of medical images with very rapid, adaptable, and accurate results.

Deng Ping Fan et al. [12] notes in this paper, automatic COVID-19 infection segmentation were observed. Methodology is developed to identify the infected region using segmentation network named Inf–Net. Semi-Inf-Net and MC is used for the segmentation of Ground Glass Opacity (GGO) and consolidation infection, they are accurately segmented. This study research will focus on integration of segmentation, quantification, and detection of lung infection. Also, it will work on multi-class infection labeling for automatic AI diagnosis.

Qingsong Yao et al. [13] describes the focus of this study was to implement label free segmentation for COVID-19. We observed that normal net methodology is quite good compare to other UAD methods and NN–net using bright pixel CT imagery. It's able to segment the COVID-19 lesion without labeling the dataset. So, it's reduced the time and complexity in manual labeling. The proposed unsupervised methodology is good but still requires a lot of development. Thus, it can segment only small part lesion more accurately.

6.3.10   Image Compression

Image compression is essential requirement in radiological imagery application. As the spatial and temporal resolution increase, the data generated is enormous, this led to a requirement of large bandwidth for communication and large data memory for storage of data arises, best alternate way is to use data compression algorithms to reduce the data size. More over lossless compression is used, this is preferred because of no loss of information while data compression, to achieve high compression ratio we need to exploit soft computing and computational intelligence algorithms.

Pradeep Kumar and Ashish Parmar [14] explain that this paper presents the lossless, lossy, and hybrid compression techniques for medical image compression and also it describes the various watermarking method and performance matrices. From literature it's concluded that hybrid techniques are more commonly used because it has a ability to compress both lossless and lossy in order to obtain the better compression. Performance parameters are computed based on their performance and efficiency and it's completely based on compression ratio.

Jing-Yu Cui et. al. [15] proposed example-based texture modeling for image compression, which is one of the current standard approaches. It uses fixed dictionary with texture samplers and vector quantizer. In past work, usually best predictor was selected based on its mean square error, but in addition to this framework it consists of the prediction residual and encoding rate. The compression quality is improved by selecting the accurate predictor. It's observed that proposed methodology performance good over JPEG and it has lower error than JPEG.

M. Moorthi and R. Amutha [16] explain that image compression mainly focuses on reducing image data, thus it's easy to store and transmit the data efficiently. Initially, segmentation is applied to obtain the two clusters i.e., region of interest and non-region of interest. Here, higher energy cluster uses the integer wavelet transform based compression and another cluster uses JPEG, which is one of the popular image formats. Finally, the designed model was able to preserve the edge information and maintain high compression ratio to provide reliable and faster compression technique.

T. G. Shrisat and V. K. Bairagi [17] note that medical images are very sensitive and it must have very clear information without any loss, it can be achieved from lossless compression. The performance enhancement in lossless compression combining predictive coding and integer transform. By looking toward the comparative analysis its predictive method gives precise compression compare to the plane wavelet-based image compression. There is very less possibility to lost a very less amount of data, by using predictive coding technique where we consider different (subtracting original image and reconstructed image) or prediction error. System performance was computed using scale entropy and entropy for compressed images with acceptable image quality.

Jiang et al. [18] explains that this paper introduces wavelet base image compression algorithm for radiological imagery, this is one the superior method for lossless image compression and improved vector quantization. The ultimate aim of proposed methodology is to maintain medical image at high compression ratio, which contains diagnostic related data. Initially wavelet transformation was applied. For low and high frequency a lossless compression method and novel vector quantization (VQ) with variable block size has been implemented respectively. Its analyzed that experimental result of optimized method can able to improve image compression performance and achieved proper balance between image visual quality and image compression ration with better performance. Proposed method was tested on liver and brain images also observed with compression ratio of 25 with different contrast ratio.

Image synthesis is the technique of generating new images from some form of image description. These images are synthesized typically like 3-dimension organ geometric shapes. Current trend requirement of medical image processing is to synthesize 3-dimension shapes. Deep neural network and capsule network with computational intelligence it is possible to synthesis any complex shapes.

Image quantification is assessing the degree of disease for a diagnosis, this is similar to computer assessing any disease without doctor interventions. This is one of the most powerful tool and future trend in radiological application, image quantification is a method to measure and classify objects as healthy or diseased. The advantage of computerized image quantification algorithm (CADe/CADx) is its objectivity and speed.

6.4   Fuzzy Logic

Fuzzy logic is a very interesting topic of AI. It allows membership value between 0 and 1, gray level and in linguistics form such a small tolerance. In traditional logic something can be represented by either having value of true, which is 1 or false, which is 0 and in fuzzy logic you can be anywhere and between 0 and 1. So you could have something that's true but only partially true and false with certain degree. Let's use tap water temperature as an example in traditional logic—you would have this represent either hot or cold representing 0 or 1, respectively. But using the fuzzy logic you could have something like this we have gradient from hot to cold so you could have something that is lukewarm very hot somewhat cold etc., instead of just cold or hot.

Fuzzy logic concept we encounter in our day-to-day life like the car is fast. The bag is heavy. Today is a hot day. Our exam was easy, and so on. None of the sentences have any metrics numbers or digits with them and yet we perfectly understand and use them every day. We are doing all of this with outlet having precise information or mathematical model of the system we can use fuzzy logic to produce accurate results in presence of inaccuracies.

An example of the fuzzy logic concept for better understanding is learning to drive a car. When you are learning to drive for the first time you are much more cautious and you create a mental rule for yourself, for example, if you are driving first and distance between your car and the car in front of you is less than a certain value you should break immediately. So, the concept of driving fast is not a fixed rule or agreed value some people tend to drive faster and then others and some people are reckless when it comes to driving it means that the concept that the driving too fast can be driving above 70 mph for a new learner and someone with a couple of year of driving experience it could be over 90 mph and for some obviously driving above 135 mph or more regardless of each group of these people are driving.

If they sense that the distance between their car and the car in front of them is shorter than they will slow down immediately even the reckless ones. So, using fuzzy logic we can design and develop systems that could drive safely regardless of the type of the driver that uses them.

Fuzzy Logic Working: Fuzzy logic is an extension of Boolean logic based on the mathematical concept of fuzzy sets. Fuzzy logic contains different components, the block diagram of basic fuzzy system is as follows shown in the Figure 6.8.

a) The fuzzifier is the part responsible for fuzzification. It's the process of converting crisp set data into fuzzy set data. It has the membership function for linguistic variable of fuzzy set.

b) Fuzzy rule-based system is an extension of fuzzy logic concept. It consists the two main components:
   - Inference engine: This process maps the fuzzy output by combining the membership function with fuzzy control rules. Fuzzy inference is the processing unit based on fuzzy set theory concept each rule having a weight between 0 and 1 and then multiplying with the membership value that is assigned to the output vector. When the input is specified the fuzzy inference, process obtain the output from the fuzzy rules-based system.

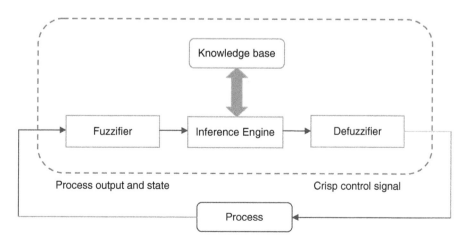

Figure 6.8 Fuzzy logic block diagram.

- Knowledge base: This is the third layer of a fuzzy system. It is the most important part of a fuzzy logic system. It is a combination of database and a rule base. It will store the knowledge available about the problem being solved in linguistic "IF-THEN" rules.

Knowledge base is constructed either by experts or self-learning algorithm.

1) The first way is for experts to construct a rule base. Experts is a system that describes if-then rules.
2) The second way is using self-learning to construct the rule base. In this method one part is used to train, while other part is to be solved by the system. These types of self-learning system are called neuro-fuzzy system.

For example: Knowledge-based system in medicine

In medical domain data is acquired from patient history, laboratory tests, physical examination, and clinical investigations. These obtained data have been converted into linguistic concept to idle medical knowledge level such as treatment recommendations, diseases description, and prognostic information.

c) Defuzzifier is the process that maps fuzzy inference output into crisp logic based on corresponding membership degrees and fuzzy set. The decision-making algorithm selects bet crisp value.

J. Greeda et Al. [19] present application of the fuzzy expert system (FES) in medicine that has been discovered to support a practioner for decision making. Fuzzy set theory plays very important role in diagnostic decision. FES used in prediction of patients conditions, patients monitoring, handling of fuzzy queries, and prediction of aneurysm, fracture healing etc. Fuzzy logic is based on human thinking decision building that produces qualitative quantitative evaluation of medical facts.

Shruti Kambalimath and Paresh C. [20] describe how fuzzy models were developed in various hydrology and water resources. The fuzzy logic-based system deal with the problems that have uncertainty, approximation vagueness, and partial truth where the data are limited, but it is not suitable for mathematical imagery and solutions because of absence of mathematical explanation. This paper suggested that hybrid fuzzy modeling is more efficient where it will be combined with ANN fuzzy SVM model to obtain better accuracy than pure fuzzy model.

Novruz Allahverdi's [21] paper describes application of FES in medical area such as determination of diseases risk and coronary heart diseases risk, periodontal dental disease, child anemia, etc. It's concluded that fuzzy control and hybrid system will give effective outcome in upcoming future. The proposed fuzzy model has been trained and tested to get proper approximation between predicted value and measured value for precise outcome.

Jinsa Kuruvilla and K. Gunavathi [22] propose FIS and ANFIS models for classification of lung cancer using CT images. Morphological operations are used to segment the lung lob from CT images and classification is done by statistical and GLCM parameters. Cluster shade, dissimilarities, skewness, and difference variance are some parameters selected by principal component analysis that is used for feature selection purpose. Adaptive neuro fuzzy inference system uses the modified training algorithm and obtained 94% classification accuracy where FIS obtained 91.4% accuracy.

Maria Augusta et al. [23] presents a fuzzy inference system created to support medical diagnoses in real time. In this paper they have done some analysis on public and private health-care services. Which describes the problems in health-care sectors are like poor allocation of resources, social inequality, Inefficiency and absence of preventive medicine. This paper has used preventive medicine as the output of the intelligent system. Using fuzzy intelligent system, they have shown the feasibility of generating new channels for medical cost. They are working to improve hospital marketing, socially responsible, to minimize wait time, cost and marketing customer contented.

6.5 Artificial Neural Network

In 1943, ANN was first time proposed by Warren McCulloch and Walter Pitts. Neural networks are inspired by human brain biological neurons; ANN are built on assemblies of connected nodes or units know as artificial neurons. Individual connection identical to synapses in a biological brain, capable of transmitting information to other neurons. These artificial neurons once receive an information process and sends to next connected neurons, the results of individual neurons are calculated by nonlinear function of the sum of its inputs. Neurons have weights they alter the learning process, these weights changes based on certain threshold. A simple neural network mainly consists input, output, and hidden layers, the quantity of hidden layer depends on the requirements, shown in Figure 6.9. ANN, finds many applications in radiological imagery, starting from prepressing to identification and classification. ANN are the best when the data dimensions are small but when the data is large, they don't perform good, due to this most of the current advanced application is radiological imaging used Deep Learning Deep Neural Networks.

ANN is a mathematical illustration of human neural architecture that reflects its "generalization" and "learning" abilities. Thus, it belongs to AI. ANN is broadly applied in research because they can model nonlinear structure where the variables relationship is unknown and very complex. ANN can have single or multiple layers. It consists of series of neurons or nodes that are interconnected in layer by using a set

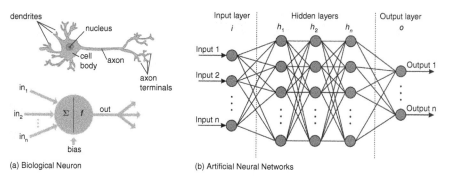

(a) Biological Neuron

(b) Artificial Neural Networks

Figure 6.9 Biological and artificial neural networks.

of adjustable weights. Each neuron is connected with each and every neuron in the next organized layer. Generally, ANN consists of three layers i.e., input layer, hidden layer, and output layer. The input layer neurons receive the information and that will be passes to the next hidden layer via weight links. Here, one or more hidden layer processed the data mathematically and try to extract the pattern. Each neuron has weighted inputs, transfer function, and single output. Neuron is activated by weighted sum of inputs it receives and activation signal process through a transfer function to produce a single output. Ultimately, last layer neuron provides final network's output.

ANN classification is in Figure 6.10, shows the multiple types of neural networks. Based on the application most suitable neural network uses with their own specifications and levels of complexity, mainly two types of ANN.

a) Feed-forward neural network

It is used more commonly in which the information is unidirectional i.e., from input to output. No feedback loops are present in this type of ANN. It's used for recognition of pattern and it contains fixed input and output.

b) Feedback ANN

In this particular ANN, it allows feedback loops. It's used by internal system error connections and used in content addressable memories.

Al-Shayea [24] reports the proposed diagnosis ANN is a powerful tool that deals with the complex clinical data and help the doctor for proper treatment of diagnosis. In this paper we analyzed the two cases, first is acute nephritis and the second is heart disease. Feed-forward back propagation network with supervised learning [1] is used as classifier in both diseases where, it's able to classify infected or non-infected person in heart diseases by 95% correctly classified while in acute nephritis, network having the abilities to learn the pattern based upon selected symptoms and proposed network were capable to classify with 99% accuracy.

Shahid et al. [25] proposed study estimated that ANN can be applied to all level of health care organizational decision making. It's found that hybrid approaches are very effective to reducing challenges such as having insufficient data or new item is introducing to the system. The most successful purposes of ANN is observed in extraordinarily complex medical situations. It's found that ANN to be often used for prediction, classification, and clinical diagnosis in area of telemedicine, organizational behavior, and cardiovascular.

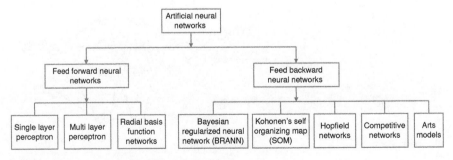

Figure 6.10 Framework for ANN classification.

Amato et al. [26] describes ANN is a powerful framework to assist physicians and other enforcement. ANN has proven suitable for various diseases and their use makes the diagnosis analysis more reliable and consequently increases patient satisfaction. This paper was describing the workflow of ANN analysis, which include, the major steps such as feature selection, database building, preprocessing, training, testing and verification for rapid and correct diagnosis prediction of various diseases.

Abiodun et al. [27] survey was present the application of neural network in real-world scenario. It's concluded that ANN can apply to any areas of industries, bio-medical, and profession fields. Based on data analysis factors it's observed that ANN is more effective, successful, and efficient. Therefore, it has ability to solve complex and non-complex real life problems. Finally result can be summarized on various fields of ANN applications regarding pattern recognition, prediction, and classification.

Mossalam and Mohamad Arafa [28] the proposed model uses the databases of 48 projects. The study aim is to identify the variables and enterprises databases that define project criticality also, the related information to build strong neural network model. The four major steps are implements to develop and test the proposed ANN model i.e., data preparation, training, testing, and sensitivity analysis. This research uses three configuration models of nets to develop intelligent model rather than existed manual selection process. Result has been observed by comparative analysis of PN method, Multi-layer feed forward network (MLFN) and best net search. Best net search generated best predictions result for the data.

6.6 Evolutionary Computation

These algorithms are very much preferred in the area where mathematical are incompatible to solve the broader range of problems and usually in the application of DNA analysis and scheduling problems, example is shown in Figure 6.11. One of the prominent evolutionary algorithms is Genetic algorithm. Below shows the procedure of genetic algorithm is these evolutionary algorithms aim in bringing out novel artificial evolutionary techniques exploiting the strength of the natural evolution and are most probably engaged in the search optimization problems that requires an optimal result.

Pena-Reyes and Moshe Sipper [29] note that the paper focused mainly on evolutionary computation (EC) and its medical application and observed the effectiveness of various evolutionary algorithms in medicine. EC makes the use of metaphor of natural evolution. EC family introduced powerful techniques that are used to search complex spaces. There are mainly three tasks which are demonstrated by EC i.e., data mining; signal processing and medical imaging; and scheduling and planning. In data mining EC work as a parameters filter. It has the ability to discover the necessary knowledge for interpretation of accumulated information. The prognosis is used frequently because of its predictive nature. EC used to performance improvement in signal processing algorithms such as compressor and filters and also from welter of data the required clinical data are extracted by taking the help of EC. It plays a very important role in planning and scheduling. Specifically, for

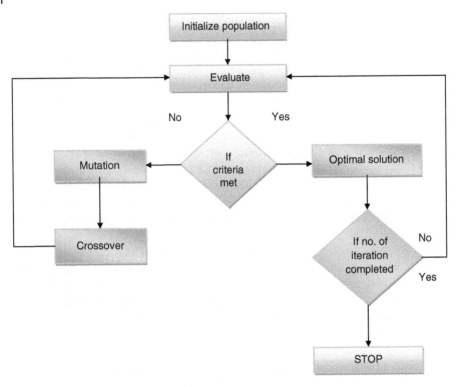

Figure 6.11 Evolutionary computation.

3-dimensional radiography, different medical procedures, and many more. EC applied in medicine in order to perform several tasks in diagnosis, especially for decision support.

Mahesh and J. Arokia Renjit [30] the actual intention of this survey paper was to review and study the segmentation and evolutionary intelligence that include various classification techniques. This review article very nicely present segmentation and classification-based approaches for the recognition of brain tumor from MRI imagery. Here, 50 research papers were studied and analysis was done accordingly, which was mainly emphasis toward the utilized image datasets, feature extraction techniques, image modality, evolution measures, implementation tools, and final achieved results. It's concluded that performance of all proposed techniques were good with respect to different modalities and its requirements, but still lots of improvement is necessary to get desire outcome. Compared to all the other existed classification-based techniques hybrid techniques gave good classification accuracy. Here, nonlinear classification and 3-dimensional evolution is complex. It has been seen that a greater number of researchers focused on classification techniques for recognition of tumor.

Nakane et al. [31], this literature review paper briefly summarized computer vision applications using evolutionary algorithms (EAs) and Swarm algorithm (SAs) and observed characteristics and differences. The proposed study concentrated on four algorithms i.e., differential evolution (DE) and genetic algorithm (GA), that belongs to EAs and colony optimization (CO) and particle swarm optimization (PSO) are belongs to SAs. Among all of the four representative algorithms of EAs and SAs, the GA and PSO are more commonly adopted in computer vision application because of its improved efficiency, parameter tuning and practical applications. Combination of the EAs/SAs and also deep neural networks such as neural architecture search is one of the popular fields of research. Ultimate aim of computer vision is to understand meaningful information and extract the features from videos and images, it's concluded that evolutionary algorithms and swarm algorithm have a potential to solve the various complex problem very precisely.

Holmes [32] evolutionary computation is a more popular approach that could be used alone or with machine learning. Proposed approach of evolutionary computation was divided as genetics based and non-genetic based algorithm. It's model free and also has a capability to provide meta-heuristic structure, where there is no need of any perfect data and prior assumptions. Thus, it can solve the wide range of problem. Genetic algorithm can be used to identify MRNA targets, also it identifies lesion on mammogram and mining temporal workflow data etc.

Slowik and Halina Kwasnicka [33] present on application of evolutionary family for real-life problems. A complete family of evolutionary optimization algorithm is considered as evolutionary computation. The paper describes the main properties of various evolutionary computation algorithms. For the easy implementation pseudo-code form is presented for each technique of EC. The described literature review gives the overview of all EC methods, which was suitable for many industrial and engineering problem, but there were some little gaps between practical and theoretical aspects. Currently EAs are modifying for hybridization of more algorithms in order to obtain better performance result.

6.7 Challenges

Tremendous improvements in image acquisition sensors has revolutionized the radiological imagery applications, over the past two decades image quality and information obtained is very large and it is helping radiologist at the same time for proper and accurate assessment of disease. Now big challenges are to develop hardware architectures that can process these data at high speed at affordable cost, there is a need to improve the processing speed.

6.8 Conclusion

This chapter provided an overview of computational intelligence application to radiological imagery. It introduced the fundamental principles of digital image processing, steps involved in image computation, and coverd computational intelligence paradigms based on fuzzy logic, ANNs, and EC. Finally, the chapter described a few applications of these paradigms and emphasized how algorithm could be made more intelligent and process at high speed with better accuracy.

Bibliography

1 Sanchez, M.G., Sánchez, M.G., Vidal, V., Verdu, G., Verdú, G., Mayo, P., and Rodenas, F. (2012). Medical image restoration with different types of noise. *2012 Annual International Conference of the IEEE Engineering in Medicine and Biology Society*, 4382–4385. IEEE.

2 Lakshmi, A. and Rakshit, S. (2010). An objective evaluation method for image restoration. *Journal of Electrical and Computer Engineering* 2010.

3 Raid, A.M., Khedr, W.M., El-Dosuky, M.A., and Aoud, M. (2014). Image restoration based on morphological operations. *International Journal of Computer Science, Engineering and Information Technology (IJCSEIT)* 4 (3): 9–21.

4 Mittal, M., Verma, A., Kaur, I., Kaur, B., Sharma, M., Goyal, L.M., Roy, S., and Kim, T.-H. (2019). An efficient edge detection approach to provide better edge connectivity for image analysis. *IEEE Access* 7: 33240–33255.

5 Hamad, Y.A., Simonov, K., and Naeem, M.B. (2018). Brain's tumor edge detection on low contrast medical images. *2018 1st Annual International Conference on Information and Sciences (AiCIS)*, 45–50. IEEE.

6 Badriyah, T., Sakinah, N., Syarif, I., and Syarif, D.R. (2019). Segmentation stroke objects based on CT scan image using thresholding method. *2019 First International Conference on Smart Technology & Urban Development (STUD)*, 1–6. IEEE.

7 Zotin, A., Simonov, K., Kurako, M., Hamad, Y., and Kirillova, S. (2018). Edge detection in MRI brain tumor images based on fuzzy C-means clustering. *Procedia Computer Science* 126: 1261–1270.

8 Cui, X., Deng, Y., Yang, G., and Wu, S. (2014). An improved image segmentation algorithm based on the watershed transform. *2014 IEEE 7th Joint International Information Technology and Artificial Intelligence Conference*, 428–431. IEEE.

9 Abubakar, F.M. (2012). A study of region-based and contour-based image segmentation. *Signal & Image Processing* 3 (6): 15.

10 Dubey, S.K. and Vijay, S. (2018). A review of image segmentation using clustering methods. *International Journal of Applied Engineering Research* 13: 2484–2489.

11 Gongwen, X., Zhijun, Z., Weihua, Y., and Li'Na, X. (2014). On medical image segmentation based on wavelet transform. *2014 Fifth International Conference on Intelligent Systems Design and Engineering Applications*, 671–674. IEEE.

12 Fan, D.-P., Zhou, T., Ji, G.-P., Zhou, Y., Chen, G., Fu, H., Shen, J., and Shao, L. (2020). Inf-net: automatic covid-19 lung infection segmentation from CT images. *IEEE Transactions on Medical Imaging* 39 (8): 2626–2637.

13 Yao, Q., Xiao, L., Liu, P., and Zhou, S. K. (2021). Label-free segmentation of covid-19 lesions in lung ct. *IEEE Transactions on Medical Imaging*.

14 Kumar, P. and Parmar, A. (2020). Versatile approaches for medical image compression: a review. *Procedia Computer Science* 167: 1380–1389.

15 Cui, J.-Y., Mathur, S., Covell, M., Kwatra, V., and Han, M. (2010). Example-based image compression. *2010 IEEE International Conference on Image Processing*, 1229–1232. IEEE.

16 Moorthi, M. and Amutha, R. (2011). An improved algorithm for medical image compression. *International Conference on Computing and Communication Systems*, 451–460. Berlin, Heidelberg: Springer.

17 Shirsat, T. G. and Bairagi, V. K. (2013). Lossless medical image compression by integer wavelet and predictive coding. *International Scholarly Research Notices* 2013.

18 Jiang, H., Ma, Z., Hu, Y., Yang, B., and Zhang, L. (2012). Medical image compression based on vector quantization with variable block sizes in wavelet domain. *Computational Intelligence and Neuroscience* 2012.

19 Greeda, J., Mageswari, A., and Nithya, R. (2018). A study on fuzzy logic and its applications in medicine. *International Journal of Pure and Applied Mathematics* 119 (16): 1515–1525.

20 Kambalimath, S. and Deka, P. C. (2020). A basic review of fuzzy logic applications in hydrology and water resources. *Applied Water Science* 10 (8): 1–14.

21 Allahverdi, N. (2014). Design of fuzzy expert systems and its applications in some medical areas. *International Journal of Applied Mathematics Electronics and Computers* 2 (1): 1–8.

22 Kuruvilla, J. and Gunavathi, K. (2015). Lung cancer classification using fuzzy logic for CT images. *International Journal of Medical Engineering and Informatics* 7 (3): 233–249.

23 de Medeiros, I. B., Machado, M. A. S., Damasceno, W. J., Caldeira, A. M., dos Santos, R. C., and da Silva Filho, J.B. (2017). A fuzzy inference system to support medical diagnosis in real time. *Procedia Computer Science* 122: 167–173.

24 Al-Shayea, Q., El-Refae, G., and Yaseen, S. (2013). Artificial neural networks for medical diagnosis using biomedical dataset. *International Journal of Behavioural and Healthcare Research 21* 4 (1): 45–63.

25 Shahid, N., Rappon, T., and Berta, W. (2019). Applications of artificial neural networks in health care organizational decision-making: a scoping review. *PloS One* 14 (2): e0212356.

26 Amato, F., López, A., Peña-Méndez, E. M., Vaňhara, P., Hampl, A., and Havel, J. (2013). Artificial neural networks in medical diagnosis. 47–58.

27 Abiodun, O. I., Jantan, A., Omolara, A. E., Dada, K. V., Mohamed, N. A., and Arshad, H. (2018). State-of-the-art in artificial neural network applications: a survey. *Heliyon* 4 (11): e00938.

28 Mossalam, A. and Arafa, M. (2018). Using artificial neural networks (ANN) in projects monitoring dashboards' formulation. *HBRC Journal* 14 (3): 385–392.

29 Pena-Reyes, C.A. and Sipper, M. (2000). Evolutionary computation in medicine: an overview. *Artificial Intelligence in Medicine* 19 (1): 1–23.

30 Mahesh, K.M. and Renjit, J.A. (2018). Evolutionary intelligence for brain tumor recognition from MRI images: a critical study and review. *Evolutionary Intelligence* 11 (1): 19–30.

31 Nakane, T., Bold, N., Sun, H., Lu, X., Akashi, T., and Zhang, C. (2020). Application of evolutionary and swarm optimization in computer vision: a literature survey. *IPSJ Transactions on Computer Vision and Applications* 12 (1): 1–34.

32 Holmes, J.H. (2014). Methods and applications of evolutionary computation in biomedicine. *Journal of Biomedical Informatics* 49 (C): 11–15.

33 Slowik, A. and Kwasnicka, H. (2020). Evolutionary algorithms and their applications to engineering problems. *Neural Computing and Applications*: 1–17.

Acknowledgment

We would like to express our gratitude to Principal, KLS GIT, Belagavi and the management of KLS society for providing opportunity to carry out research in association with JN Medical College, KLE Academy of Higher Education and Research (Deemed-to-be-University), Belagavi.

Dr. Hari Prabhat Gupta (SMIEEE, hprabhatgupta@gmail.com, https://sites. google.com/site/hprabhatgupta) is an Assistant Professor in the Department of Computer Science and Engineering, Indian Institute of Technology (BHU) Varanasi, INDIA. Previously, he was a Technical Lead in Samsung R&D Bangalore, India. He received his Ph.D. and M.Tech. degrees in Computer Science and Engineering from Indian Institute of Technology Guwahati in 2014 and 2010 respectively; and his B.E. degree in Computer Science and Engineering from Govt. Engineering College Ajmer, India. His research interests include the Internet of things (IoT), Wireless Sensor Networks (WSN), and Human-Computer Interaction (HCI). Dr. Gupta got various awards such as Samsung Spot Award for outstanding contribution in research, IBM GMC project competition, and TCS Research Fellowship. He has guided 3 Ph.D. thesis and 5 M.Tech. dissertations. He has completed two sponsored projects. He has published three patients and more than 100 IEEE Journal and conference papers.

Swati Chopade (swatischopade.rs.cse18@itbhu.ac.in) received M.Tech Degree in Computer Science and Engineering from VJTI, Mumbai, India. Presently, she is pursuing Ph.D in Department of Computer Science and Engineering, IIT (BHU) Varanasi. Her research interests include machine learning, sensor networks, and cloud computing.

Dr. Tanima Dutta (SMIEEE, dutta.tanima@gmail.com, https://sites.google.com/ site/drtanimadutta) is an Assistant Professor in the Department of Computer Science and Engineering, Indian Institute of Technology (Banaras Hindu University), Varanasi, India. Previously, she was a Researcher in TCS Research & Innovation, Bangalore, India. She received Ph.D. in Dept. of Computer Science and Engineering, Indian Institute of Technology (IIT) Guwahati in 2014. Her Ph.D. was supported by TCS (Tata Consultancy Services) Research Fellowship and she received SAIL (Steel Authority of India Limited) Undergraduate Scholarship for perusing her B.Tech. Degree. Her research interests include (MAJOR) Deep Neural Networks, Machine Learning, Computer Vision, and Image Forensics and (MINOR) Human-Computer Interaction (HCI) and Intelligent Internet of Things (IIoT).

7

Computational Intelligence in Agriculture

Hari Prabhat Gupta[1], Swati Chopade[2] and Tanima Dutta[3]

[1] Assistant Professor in the Department of Computer Science and Engineering, Indian Institute of Technology (BHU) Varanasi, India
[2] M.Tech Degree in Computer Science and Engineering from VJTI, Mumbai, India
[3] Assistant Professor in the Department of Computer Science and Engineering, Indian Institute of Technology (Banaras Hindu University), Varanasi, India

7.1 Introduction

The prime purpose of agriculture is to produce foods, vegetables, and crops by cultivating the land with natural environmental resources, such as the soil, water, etc. Food is one of the essential need for living beings to sustain life [1]. Due to climate change, problems such as crop diseases, lack of storage management, etc. can become a bottleneck to attain high crop yield. Nowadays, the maximum production of foods from the available land is necessary for the rapidly growing population. There exists low-power, highly-portable, and low-cost agricultural sensors to monitor the land for increasing crop production. Examples of agricultural sensors are temperature and humidity sensor, green sensor, camera sensor, environmental sensor, and soil moisture sensor, etc. The green sensor sense the reflectance of wavelengths for a green light to give information about chlorophyll in the leaves, evaluating Nitrogen (N) status present in the leaf. The soil moisture sensor informs about the dampness level of the soil. Next, the temperature and humidity sensor (DHT11) senses the temperature and humidity of the surrounding environment. The camera sensor senses the light and converts it into signals resulting in the formation of an image. Further, the environmental sensors sense, measure, monitor, and record the environmental parameters such as humidity, heat losses, etc. Figure 7.1 shows the sensors utilized in modern agriculture for sensing and monitoring the field. Integration of sensor technologies with the existing agriculture has given rise to a revolutionary transformation in agriculture. With the use of sensor technology, the Global System for Communications (GSM), Global Positioning System (GPS), control systems (such as cooling fan, heater, pump motor), quadcopters (or drones) and software services to find the optimal soil moisture content conditions, fertilizers, and crops have become possible. For example, Unmanned Aerial Vehicle (UAV) helps in agricultural data collection to facilitate the easy assessment of the environmental factors, crop features and soil conditions. This is beneficial to the farmers. Due to easy and convenient sensors, smart sensing strengthen the sensing and controlling the agriculture remotely without human intervention through a network model [3]. The sensors collect real-time agricultural data related to the soil, crops, and surrounding environmental factors. The farmers can manage the agricultural activities by using the predictive analysis of the collected data that helps them decide in advance.

Emerging Computing Paradigms: Principles, Advances and Applications, First Edition.
Edited by Umang Singh, San Murugesan and Ashish Seth.
© 2022 John Wiley & Sons Ltd. Published 2022 by John Wiley & Sons Ltd.

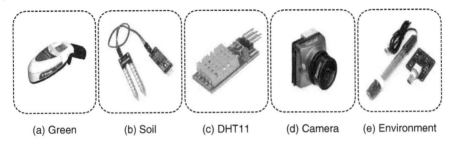

(a) Green   (b) Soil   (c) DHT11   (d) Camera   (e) Environment

Figure 7.1 Agricultural sensors for collection of sensory data [2]. *Source:* Image Source for Agricultural Sensors, 2021. [Online]. Available: https://www.google.com/search?q=agricultural+sensors; Sharvi Electronics; RunCam Technology Co., Ltd.

Computer-based systems and automation are becoming essential parts of modern agriculture. There exists a vast number of challenges where the use of traditional methods are not sufficient to handle. Examples of such challenges are the need for food increases briskly due to rapid population growth, or the extensive use of pesticides can hamper the soil resulting in no fertility to enhance the productivity, etc. The various complex tasks formerly performed by humans are now taking over by computer-based systems with superior performance. Thus, there is a desperate need of sophisticated methodologies to handle the increasing complexity of the existing agricultural systems. This leads to the creation of a new emerging field of Computational Intelligence for Agriculture (CIA). The CIA is a soft computing paradigm to improve the productivity and quality of the crops. It includes different methodologies that are very close to the human's reasoning and decision making to facilitate automation and minimize human efforts. The CIA incorporates a highly portable and small in size agricultural sensors for monitoring the various architectural field activities. These sensors generate significant and real sensory data utilized by the new emerging technologies for automation practices such as Machine Learning (ML) and Deep Learning (DL) to train the various prediction models of the CIA. Due to these technologies, nowadays, the CIA is becoming more popular for agriculture. Further, the CIA uses ML and DL techniques to automate multiple processes such as monitoring the soil and plant health, the growth rate of crops, detecting abnormalities in crop growth, improving productivity and quality, etc. The significant difference between these two technologies of CIA is ML utilizes handcrafted features obtained by using statistical operations such as standard deviation, average minimum, etc. and DL is capable of extracting explanatory features automatically from the input data by performing enormous computations to find complex and hierarchical patterns from real-world problems.

7.1.1 Machine Learning (ML) Models for CIA

ML is an application of Computational Intelligence (CI) that teaches computer-based programs to think similarly to how humans do, learning and enhancing upon past experiences. Thus, it gives the capability to learn on its own and correct from experience or historical data without an explicit program. ML concentrates on the buildup of computer programs that can access the multi-dimensional and multi-variety data, explore the data, identify patterns in the data, and involves minimal human intervention. With the help of ML techniques such as multiple linear regression, K-means clustering, and decision trees, etc., it can build a model based on accessed accurate and extensive data resulting in the trained model, capable enough to make decisions for agriculture [4]. ML consists of different forms of data or features, such as numeric (integer, real, number, etc.), nominal data such as various crops (wheat, corn, soybean), binary (0 or 1), or ordinal data such as fertilizer status (less, accurate, more). To measure ML models' performance, the performance metric, capable enough to improve with experience over time, such as recall, precision, and F1 score (considering precision and recall simultaneously), etc. utilized. After the completion of the learning process, it is necessary to use the trained model to categorize the new testing data with more accurate results. ML tasks broadly categorized into three main classes, namely, supervised learning, unsupervised learning, and reward-based learning or semi-supervised learning. There are input variables and output variables in supervised learning, and the model learns the mapping function from input to output by using the training data. Figure 7.2 illustrates the ML classifier for agricultural data. Further, the supervised learning problems can be categorized into classification and regression problems based on the given task. In classification problems, the output variable is a categorical variable, for example, "yes" or "no," and the output variable is a real value (*floating point number*) in regression problems such as "dollar" or "height."

Support Vector Machine (SVM) is a supervised ML that uses a linear function or polynomial function, known as, kernel function to find the nonlinear function on the input data having noise. Thus, it helps in transforming the complex problems into a simple linear function optimization problem to take accurate decision helpful to the farmers. Random Forest (RF) is another supervised ML that creates multiple trees for

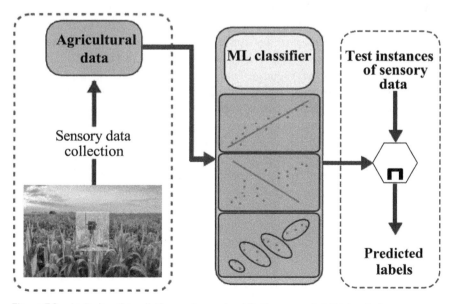

Figure 7.2 Agricultural prediction system using ML. *Source:* somkak/Adobe stock.

different features of agricultural input data and the output is decided by using the average value of the prediction of the individual tree. In unsupervised learning, there are input variables but no corresponding output variable. In other words, the labels are not available to guide the learning process. The model has to discover hidden patterns from the training data. The teacher does not guide the process of learning. The model can find a suitable structure from the distribution of input data without guidance. Next, unsupervised learning problems classified into clustering problems and association problems depending on the real-world issues. A clustering algorithm is an unsupervised ML that helps make a group of similar data points from the enormous data without concerning the output. For example, a grouping of the customers having similar purchasing behavior. Next, the association algorithm finds a general rule on the massive amount of data that applies to a large portion of input data. For example, different crops need different fertilizers. Semi-supervised learning includes vast input data, but the corresponding output data or label is available only for some input data. Thus, making its need to solve such problems by using both supervised and unsupervised algorithms. ML algorithms do not need any assumptions on the distribution of data toward classification. In other words, the unbalanced data instances can be present in different classes based on labels; still, ML is capable of making accurate categorical predictions. These techniques are capable enough to combine the different sources' data even if the data is having unknown probability density functions [5].

7.1.2 Deep Learning (DL) Models for CIA

DL, another emerging technology of CI, expands ML and imitates the workings of the human brain in processing actual data and creating patterns by analyzing the data to utilize in decision making for agriculture with promising results and with immense potential. DL incorporates multiple layers to extract higher-level features of the field progressively from the raw input data. The data is represented hierarchically with the help of different functions. It also comprises different components, namely, convolutions (use of filters on input data), pooling layers (downsampling of input data), fully connected layers (Feedforward network), gates (acts as threshold), memory cells, activation functions (for activation of neuron), encode/decode schemes. DL techniques such as Convolutional Neural Networks (CNN), Long Short-Term Memory (LSTM), and Recurrent Neural Networks (RNN) have been widely applied in various complex problems of agriculture with a considerable improvement in the performance by comparing with ML. DL algorithms generally termed data-hungry as they work excellently with vast volumes of data. The CNN model takes agricultural input data, assigns weights and biases (i.e., decides importance), and learns automatically and adaptively the spatial hierarchies of features from the farming input data with the help of back propagation and the explanatory features are identified to take proper decisions for agriculture [6]. Figure 7.3 illustrates the DL classifier for agricultural data. RNNs are capable of handling sequential agricultural data, can memorize previous inputs due to internal memory storage and can accept current input to provide the useful features for decision making. LSTM utilizes raw sensory agricultural data as a input to the sequential combination of cells. The output of the last cell is provided as an input to the fully connected layer and ouput of this layer is utilized to predict the field's decision. Deep neural networks use only fully connected layers for

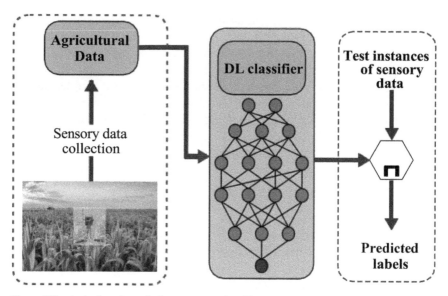

Figure 7.3 Agricultural prediction system using DL. *Source:* somkak/Adobe stock.

finding the predictions. Prediction of agricultural production output by using CIA becomes very useful to manage the post harvesting enoromous activities, namely, storage, processing, packaging, transportation, and marketing.

7.1.3 Wireless Communication Protocols for CIA

An ample of heterogeneous battery-operated sensors of CIA produces extensive sensory data. The data is further transmits to the nearest processing unit or the network server using suitable short-range or long-range communication protocol. The selection of the appropriate communication protocol is based on the size of transmission data and the distance between the sender and the receiver. The raw data is preprocessed by using different techniques, converted into a standard scale and then utilized by ML and DL to take the correct decision for agriculture. The examples of communication protocols are Long Range Wireless Area Network (LoRaWAN), Bluetooth Low Energy (BLE), Wi-Fi, etc. The description of communication protocols utilized in CIA are given in Table 7.1. The CIA techniques, such

Table 7.1 Different wireless communication protocols for CIA.

Technology	Max Coverage	Frequency	Communication	Data Rate
NFC	10 cm	13.56 MHz	Two way	200 Kbps
RFID	3 m	Varies	One Way	Varies
Bluetooth	100 m	2.45 GHz	Two Way	22 Mbps
Wi-Fi	100 m	5 GHz	Two Way	144 Mbps
Zigbee	10 m	2.4 GHz	Two Way	250 µbps
LoRaWAN	10 km	865 MHz	Two Way	300 Kbps

as ML and DL, can work with the nonlinear inexact and incomplete data to decide proper actions beneficial to the farmers. DL is more potent for the automatic extraction of defining features from sensory data of the agriculture field. For example, in the case of plant disease detection, DL can produce higher prediction accuracy for the disease. It is also helpful to broaden the scope of detected disease as well as consider different plant species also. In general, the farmers decide the crop yield by using prior experience. Due to the drastic change in weather conditions, their guess is not that correct, collapsing their financial situation. Nowadays, there is a lot of advancement in technology for agriculture. Therefore, they can take the help of technology to predict the yield. In this chapter, we cover the use of computational intelligence for agriculture. We mainly cover different topics related to CIA.

7.2 Estimation and Improvement in Crop Yield

In this section, we discuss different ML and DL algorithms used in modern agriculture to automate the task of estimating the accurate crop yield. Accurate crop yield estimation before harvest is crucial in agriculture because it helps crop management, business decisions. Enhancing crop yield production with quality (such as size, shape, taste, and nutritional value of crop) and, at the same time, reducing operational cost and environmental impact is the prime motivation of modern agriculture. The capacity for growth and yield from the land depends on various aspects such as the climate, soil conditions, region of the field, irrigation, fertilizers, etc. Nowadays, remote sensing (RS) technologies are widely available for accurate sensing of different agricultural field attributes to build a decision support system for maximizing yield production [7]. These techniques, such as multi-spectral scanning and satellite, enable collecting spatial, temporal, and spectral data about the agricultural field. The field has different crops, namely, wheat, maize, trees, vineyards, etc. and collected crops' data through satellites, UAVs, and ground-based vehicles. Figure 7.4 illustrates the modern agriculture having drone, satellite, and Unmanned Ground Vehicle (UGV) for remote sensing and forming a wireless network for communication of sensory data. However, Remote Sensing (RS) based techniques collect comprehensive sensory data from diverse platforms. The volume of the data is continuously increasing. It is beyond human capability to analyze, combine, and make a correct decision (specifically when the data is heterogeneous collected from various sensor modalities). Therefore, there is a need for ML and DL techniques to process various inputs for decision making and to incorporate expert knowledge in the modern agricultural system. ML and DL models' main task is to identify the yield-limiting factors by learning through the input parameters and help farmers solve the problems for crop growth and improve crop yield. These models are capable enough to update as soon as the new agricultural data becomes available and handle uncertainties in real-world problems using probabilities. Next, these models can process the data collected from multiple sensors. If the data instances are available for all classes (no bias toward one of the class), higher information gain is possible by using these models. Various ML models used for estimation of different crop yield in modern agriculture.

  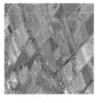 

(a) Drone   (b) UGV   (c) Satellite   (d) Wireless Network

Figure 7.4 Modern agriculture with remote sensing and wireless network [8]. *Source:* photolink/Adobe Stock Photos; Unknown author/public domain; vaalaa/Adobe Stock; Image provided by Satellite Imaging Corporation; onephoto/Adobe Stock Photos.

ML algorithms, such as Boosted Regression Trees (BRT) and SVM predicted the regional winter wheat yield by using Normalized Difference Vegetation Indices (NDVI) (derived predictors or features). Vegetation Index (VI) is a ratio (mathematical combination) of green, red, and infrared spectral bands. ML models handle well with a large set of predictors such as single NDVI, incremental NDVI, and targeted NDVI to predict the crop yield if the training samples are limited. These models used functional relationships between remote-sensing observations and crop characteristics to estimate maximum accurate crop yield [9]. Next, Gradient Boosting (GB), the conventional ML technique, is used to solve the real-world problems by using regression and classification. The GB method generated the prediction model for agricultural yield by ensembling the weak decision trees to predict the crop yield. Another conventional ML technique, known as eXtreme Gradient Boosting or X-Gboost (XGB), is used for any problems (such as classification, regression, etc.) to speed up the process, enhance the performance, and produce accurate information for yield. Some ML techniques, namely, Gaussian Process (GP), Indian Buffet Process (IBP), and Dirichlet Process (DP) are probabilistic models and consider the noisy sensory data while combining the information from different sensors to predict the yield. They used confidence intervals for predictions of crop yield. Moreover, the supervised ML models determined the future price of the product using crop yield. These models utilized the influential factors (independent variables or agricultural input parameters) and the corresponding price (dependent variable) to learn the mapping function for forecasting future prices. There are two variations of these forecasting techniques: univariate models having only one independent variable and another variation contains more than one independent variables, known as, multivariate models.

Besides, ML techniques, such as Extremely Randomized Trees (ERT), SVM, RF, and DL, estimated the yield for the corn crop providing a monetary judgement to the farmers. In general, ML algorithms, namely, Classification, Regression, RF, and Support Vector Machine in the agriculture field, considered crop yield as the dependent variable and other features such as temperature, rainfall, nitrogen status, and soil type from the input data are independent variables. These models identified correlations between the agricultural input parameters, discovered patterns from huge sensory data and found the mapping function from input to output. This mapping function helped to tell the crop yield by learning from the past historical agricultural data. Next, the generated prediction model of an accurate crop yield helps farmers decide when to grow and what to grow in advance. Further, to handle complex agricultural input data and build a tree from top to bottom, Regression Trees

(RT) ML algorithms have been given better performance. Also, Boosted Regression Trees (BRT) and RF predicted accurate early-season wheat yield in the winter season using climatic data, fertilization levels, and NDVI (collected from SPOT-VEGETATION sensor) [10]. With the help of technologies, farmers can produce incredible food from the same land. Figure 7.5 illustrates the uses of various technologies such as precision agriculture and satellite images to collect an enormous sensory data of agricultural input parameters for getting a massive crop yield.

In general, the ML techniques are data-driven. Therefore, ML algorithms' exact calculation of the crop yield heavily depends on massive sensory data and relationships between input and output parameters in the sizeable sensory dataset. Next, data with noise, incomplete datasets, and the presence of biases and outliers in the data can reduce the models' predictive capability. Traditional ML models incorporate heuristics (private knowledge of the problem) to extract the prime features from input data, not able to provide a general solution. In contrast, DL models use auto-encoder (learn from compressed input) to extract the features automatically. They structure algorithms in multiple layers to make clever decisions on their own. These models have less dependency on the input data, overcomes the fundamental limitation of ML. Therefore, they perform good quality crop yield estimation even if in the presence of limited data acquisition. Various DL models used for analysis of diverse crop yield in modern agriculture are as below.

DL algorithms such as CNN and LSTM utilized different crop features such as crop itself, weight, growth of the crop, crop density, various species of plants, and a group made by combining elements to predict crop yield. Next, the LSTM DL model processed parallel agricultural input data series to generate output for taking proper agrarian decision. Despite fewer input variables, the LSTM-CNN DL model has been given accurate results by training the model using temporal structure. The model utilized three types of layers: convolutional layer, pooling layer, and fully-connected layer, to learn from input variables of agriculture data. The model has a memory cell with a computational unit providing better decisions to the farmers to maximize the crop yield. Further, the compound DL model, namely, the Attention-CNN-LSTM model, predicted the actual strawberry yield before harvest and also indicated the strawberry price before five weeks by using various parameters, namely, weather, soil, irrigation technology, environment, and strawberry yield. Thus, it protects the farmers from price crash since the prediction of the product's price for farmers is essential in the procurement process of fresh produce supply chain management [11].

DL models, such as Supervised Self-Organizing Maps (SOM), Supervised Kohonen Networks (SKN), XY-fused Networks (XY-F) and Counter-Propagation Artificial

(a) Incredible Yield.   (b) Exact Water Supply.   (c) Precision Field.   (d) Tremendous Food.

Figure 7.5 Various uses of technology to improve crop yield [8]. *Source:* Vijayanarasimha / 247 images / Pixabay; Deyan Georgiev/Adobe Stock Photos; zapp2photo/Adobe Stock; The Science and Information (SAI) Organization Limited; kinwun/Adobe Stock Photos.

Neural Networks (CP-ANN), determined the yield of wheat by using fusion vectors of eight soil data parameters, crop growth characteristics (NDVI values) from satellite images and previous years' yield data. These models used proximal online soil sensing to find within-field variation and estimated soil properties by minimizing the time and labor cost of soil sampling and analysis [12]. Next, despite limited data acquisition, DL model such as Spiking Neural Network (SNN) predicted land use or crop production by automatically extracting defining features from agriculture input data. Further, the Deep Gaussian Process (DGP) predicted country-level soybean production by integrating spatial and temporal information from the agricultural data beneficial to the country.

7.3 Water Conservation and Environmental Change

Here, we discuss water management in the field of agriculture. Water is the most valuable environmental resource for the development of crops. Water conservation for sustainable agriculture is efficient water management to make it available when and where required for crops. It is the combined process of evaporation and transpiration. The evaporation converts the water into vapor or gas, and in the transpiration process, the plants get the water through their roots and give off water vapor through their leaves. It is helpful for the design and management of irrigation systems for agriculture. Figure 7.6 illustrates the different water conservation techniques for agriculture. Rainwater harvesting refers to collect the rainwater and store it for later use. Next, in drip irrigation, the water drips slowly to the plant's roots, resulting in saving water. Further, a series of pipes and sprinklers coupled together to form a lateral movable system to water the plant. Finally, a channel is to carry water from the river or reservoir to the field. Another factor affecting crop yield is the changing environment. The environmental change includes climate change, drought, flood, freshwater shortages having an impact on the growth of crops. Climate refers to the weather for a region over a long period producing vulnerability to the field. The climatic components affecting the agricultural productivity are CO_2, precipitation (such as rain, ice, hail), wind speed, soil moisture, etc. Environmental change creates diseases in the plants, reduction in soil moisture, and prevents crops' improvement. With ML and DL's help, it is possible to forecast the weather and take appropriate actions well in advance to protect the crops and increase the production of crops. Various ML and DL models used in estimation for water conservation and how to handle environmental change in modern agriculture.

(a) Rain Harvesting   (b) Lateral System.   (c) Drip Irrigation.   (d) Channel Creation.

Figure 7.6 Water conservation for agriculture [8]. *Source:* Krzysztof Dudzik/Wikimedia Commons; Cecilia Lim/Adobe Stock Photos; Pichit/Adobe Stock; Indian Council of Agricultural Research; Biman Saha; Samuel Mederos/Adobe Stock; Emoji Smileys People/Adobe Stock Photos.

ML algorithms estimated evapotranspiration to decide practical usage of irrigation systems and developed a cost-effective intelligent irrigation system to water the plant automatically without wastage of water with low-cost sensors. The ML and DL models identified expected weather phenomena or environmental change by predicting daily temperature, and helped farmers make appropriate decisions. Next, the CNN and SVM models determined the stress available in the soybean leaf due to environmental change and prevented productivity loss and enhanced the financial condition of the farmer. Further, Linear Regression (LR) algorithm forecasted the powdery mildew disease in the wheat crop using weather data, such as solar radiation, precipitation, humidity, temperature, hyperspectral data such as the reflectance of the red band and variations in time for estimating the probability of disease in crops. Moreover, the ANN ML model found the relationship between weather parameters such as minimum humidity, maximum temperature, minimum temperature, maximum moisture, and rainfall responsible for diseases in the crops and predicted the illness and its severity accurately using Late Blight disease's early symptoms in potato. Next, the rule-based Artificial Intelligence (AI) or ML model used environmental parameters, such as soil humidity and its temperature, air temperature and humidity, and light intensity to estimate the exact amount of water for rinsing the green-house's organic vegetable crops automatically without the help of the human being. Figure 7.7 illustrates the different environmental changes some may be sudden such as flood due to heavy raining, infections in the fruits, drought and global warming generates the loss of productivity in the agricultural field.

ANN ML model estimated daily evapotranspiration using maximum temperature, minimum temperature, sunshine duration, speed of wind available at 10 m height, air temperature present at 2 m height, mean relative humidity accurately. The accurate calculation of evapotranspiration is helpful to the farmers to decide when to water the plant to maximize production [13]. Next, the model predicted daily dew point temperature using average air temperature, atmospheric pressure, horizontal global solar radiation, relative humidity, and vapor pressure. This prediction is helpful for water management to improve crop yield [14]. Further, the model estimated monthly average daily solar radiation using meteorological parameters, geographical parameters, and month of the year as input and generated solar radiation's quantitative output for different periods helpful farmers to make decisions about watering the plant. The model incorporated climatic

(a) Flood. (b) Disease in fruit. (c) Drought (d) Global Warming

Figure 7.7 Environmental change on the crop [8]. *Source:* Gina Sanders/Adobe Stock Photos; Scot Nelson / Flickr / Public domain; rostyle/Adobe Stock Photos; ParabolStudio/Shutterstock.com; ParabolStudio/Shutterstock.com.

parameters such as the mean of diffuse radiation, mean of beam radiation, etc. and geographical parameters such as latitude, longitude, and altitude to predict solar radiations. Moreover, the feedforward ANN ML model estimated the monthly average daily solar radiation using multilinear regression algorithm. The algorithm utilized the independent agricultural input parameters, such as sunshine duration, maximum atmospheric temperature, latitude, the month of the year, soil temperature, altitude, minimum atmospheric temperature, cloudiness, mean atmospheric temperature, and wind speed to predict solar radiations to schedule the water time table.

7.4 Soil and Plant Health

In this section, we describe the use of ML and DL for finding soil and plant health to maximize crop yield. Soil is a natural environmental resource having vital nutrients for plants. In other words, it is an untied material consisting of mineral, air, tiny particles of rocks, water, and available on the land's top. Figure 7.8 illustrates the variations of soil for the plant's growth, such as healthy soil, the soil having nutrients used as medicine for plants, lousy soil, and good soil for plant growth. Healthy soil is capable enough for water storage and filtration, carbon capture and storage, biological function diversity, and productive capacity. There are many ways, such as mulching, tillage operation, conservation of tillage, and crop rotation to maintain or improve the health of the soil and keeping the soil health ultimately improves plant health. Figure 7.9 illustrates the different ways to maintain the soil health for the plant's growth, such as mulch, tillage, conservation of tillage, and crop rotation. A mulch is a layer of material applied on the soil's surface that is beneficial for many reasons, such as soil moisture conservation, curtailing the weed's growth, maintaining the health of the field, and increasing fertility. Tillage operation refers to the mechanical manipulation of the soil, and conservation of tillage represents the seedbed preparation helpful for maintaining soil temperature up to a certain depth, soil conservation, soil erosion, and reducing the costs of soil preparation. Rotation of crops is beneficial to enhance soil health and increase the productivity of the farm over time. For example, the farmer can select bean plant after corn plant because beans add N to the soil, which utilized by the corn plant. Various ML and DL models used in estimation for crop and plant health in modern agriculture.

(a) Healthy soil   (b) Soil as medicine   (c) Lousy soil   (d) Plant growth

Figure 7.8 Soil and plant health [8]. *Source:* Vitaliy Sinkevich/Adobe Stock; Dmytro Smaglov/Adobe Stock Photos; diyanadimitrova/Adobe Stock Photos; maxsattana/Adobe Stock Photos.

(a) Plastic mulch   (b) Tillage operation   (c) Tillage conservation   (d) Crop rotation

Figure 7.9 Different techniques for improving soil health [8]. *Source:* zlikovec/Adobe Stock; ompstock/Adobe Stock Photos; erdemkose/Adobe Stock; UC Division of Agriculture and Natural Resources; Dusan Kostic/Adobe Stock Photos; svetamart/Adobe Stock Photos; Sunny Forest/Adobe Stock Photos.

Soil is not a homogeneous natural resource, includes very complicated processes, and the temperature of soil alone is capable enough to provide information about the climate change effects on crop yield [15]. The ML and DL algorithms utilized evaporation processes, moisture level, and temperature of the soil to understand the characteristics of the soil as well as identify intruder's entry into agriculture. The model created by using ML and DL makes use of different features, such as nutrients in the soil, pH value, area of production, nitrogen, potassium, sulphur, manganese, phosphorus, cation exchange capacity, boron, soil type, magnesium, calcium, and zinc for deciding soil health. With the descriptive ML model's help, the knowledge obtained for the leaf area index feature tells about the crop's growth [16]. Next, the ML model helped to find a mapping from soil properties to crop yield allows the farmers to locate areas of concerns and enable the implementation of site-specific field practices at a local scale. Several ML models, such as bagged trees, Gaussian kernel-based support vector machines (SVM), weighted K-nearest neighbors (K-NN), predicted the different soil categories (Barisal, Sara, Isshwardi in Bangladesh) by using chemical features (pH, Boron, Sulphur, Manganese, etc.) with land types and recommended the appropriate crops by considering geographical attributes. Further, the ML models such as RF, BayesNet, Naïve Bayes predicted the soil type beneficial to the farmers for deciding the kind of crop.

The CNN DL model used to incorporate contextual or surrounding information for the agricultural land using multiple spatial contextual (3D-stack of images) images as input, analyzed contextual information, and determined neighboring pixels' nonlinear spatial relationships. That relationships further used to manage agricultural land and predicted the soil organic carbon content present at the deeper layer of soil in the field. Moreover, the Boosted Regression Tree (BRT) ML model used the input parameters, such as topographic position indices (TPI), precipitation, red wavelengths reflectance, and evapotranspiration to predict the soil moisture content present in the soil's upper layer (having 4 cm depth). Then, that information utilized for different field and environmental related tasks modeling where the soil moisture content level is explicitly essential. Next, Generalized Regression Neural Network (GRNN) ML algorithm used to predict soil moisture level (depth is 5 cm from the surface), an essential parameter for the water cycle and global water budget. Further, RF ML model estimated soil moisture of the land surface using Cyclone Global Navigation Satellite System (CYGNSS) measurements for interpreting land-atmosphere interactions of carbon-fluxes, terrestrial water and energy at the land-atmosphere interface. The model learned the relationship between soil moisture and CYGNSS signals by considering different land parameters to get the information about soil moisture up to 9 km depth from the surface and 1−2 days temporal data useful for agriculture.

7.5 Fertilization and Pesticide

In this section, we elaborate on the use of ML and DL models for farm management by considering different fertilizers for the growth of a crop. Fertilization is an effective way to enhance the yield and improve the quality. Fertilizer is a natural or synthetic (chemical) substance added to the soil to make the plants grow better to maximize the productions. In general, plants need many components for improving their growth, out of which some are related to the soil, and the remaining three are primary elements, namely, Nitrogen (N), Potassium (K), and Phosphorous (P) required in the highest quantities, known as NPK fertilizers. N fertilizer is helpful for the growth of leaf. Further, P fertilizer speeds up the development of fruits, roots, seeds, and flowers. K fertilizer is responsible for the stem's strong growth, water movement in the plants, and advancement in fruiting and flowering. Figure 7.10 illustatres an example scenario of fertilization, injection of fertilizers and spread of pesticides in the field.

Modern agriculture incorporates chemicals in terms of fertilizers and pesticides, that are an inseparable part of it. Due to the continuous cultivation of the land, there is the possibility of a lack of nutrients. So, it is mandatory to supply these fertilizers regularly to enhance crop yield. Next, pesticides are a mixture of chemicals used to kill or control a pest in the agricultural field. A pest can be insects, fungi, weeds (unwanted plants), rodents (rats and mice), microbes (bacteria or viruses). Due to the plants' diseases, there is the crop-loss over the world, resulting in a threat to food security. Therefore to control infection in the plant, chemical pesticides are needed. With the help of the external supply of plant nutrients in the form of fertilizers and pesticides, it has become possible to achieve stability in productivity by managing

   (a) Fertilization   (b) Pesticiding   (c) Fertilizer injection   (d) Pesticiding spray

Figure 7.10   An example scenarion of fertilization and pestciding [8]. *Source:* Africa Studio/ Adobe Stock Photos; Budimir Jevtic/Adobe Stock Photos; poomsak / Adobe Stock; Arbormax Tree Service; Sergey_Siberia88/Adobe Stock Photos.

pests. Heavy use of fertilizers can deplete the nutrients and the quality of the soil, and high usage of pesticides can harm plants and sometimes human-beings also. So, with the help of ML and DL models, the farmers can decide the correct amount of fertilizers and pesticides required for agriculture and getting a cost-effective solution. The estimation of fertilizers, such as Nitrogen (N) is essential because N is considered a vital mineral nutrient for plant growth and improvement. N's optimization for different crops is a prime problem since N has an enormous environmental and significant economic impact. ML algorithms solved the optimization problem, as it can help to discover a pattern from heterogeneous data i.e., sensed by sensors with various spectral, spatial, and temporal modalities for deciding the value of N. Next, the Support Vector Machine ML model is used to quantify Nitrogen status, which gives information of N to the farmer to decide the health of the soil. The field management model helps farmers adjust their field by considering the features such as fertilization, irrigation, and soil nutrients. Thus, farm management has become convenient by knowing when to apply fertilization and pesticides to the field. Further, ML models, namely, Gaussian Process (GP) regression algorithms, estimated the N-content, specific leaf area, chlorophyll-content, leaf water content from a multi-species field-based dataset that helps farmers to take actions priorly to increase agricultural productivity.

Moreover, Cubist and Least Squares Support Vector Machines (LS-SVM), ML models predicted three different properties of fresh wet soil online, namely, the Total amount of Nitrogen available in the soil, Organic Carbon present in the soil and soil Moisture Content. Thus, the farmers can decide further actions based on these predictions [17]. Besides, Principal Components Regression (PCR), Partial Least Squares Regression (PLSR), Stepwise Multiple Linear Regression (SMLR), and Back Propagation Neural Network (BPNN) ML and DL models predicted the leaf N concentration for N fertilizer management in pear orchard fruits to maximize fruit production [18]. Next, Stepwise Linear Regression (SLR), RF, Generalized Additive Mixed Model (GAMM), Classification and Regression Tree (CART) models predicted the Total N in the soil for the rubber plantation. The total N in soil plays a vital role in soil fertility and nutrient cycling. These ML models used fourteen different environmental input parameters, namely, mean precipitation, mean temperature, topographic wetness index, parent materials, elevation, slope, aspect, mean normalized difference vegetation index, horizontal curvature, relief, stream power index, profile curvature, relative positional index, and convergence index to predict total N in the soil [19]. Further, RF ML algorithm accurately predicted N concentration present in the leaf of sugarcane using hyperspectral data. Also, RF ML model used different input parameters such as pH, type, and nutrient of the soil in that region, the concentration of N, P, and K in the soil, weather, temperature, rainfall in that region and soil composition to determine the accurate yield as well as which crop expected to give better yield in that region. Further, yield estimations predict the ratio of fertilizers required for the selected crop to get profit and sustainability. Despite scarce data, the Bayesian Network-Based Transfer Learning approach is beneficial for fertility grading (quality or level) of soil and further the transfer learning model decided the fertility quality of another agricultural land by using geographical information.

7.6 Early Detection of the Plant Diseases

In this section, we elaborate on how to use ML and DL for plant disease detection. Early detection and management of plant disease problems can help control the disease by spreading pesticides and increasing subsequent profit. Diseases due to fungal, bacterial, and viral infections because of insects, rust, etc. affect the crops' growth and productivity, and once the the disease starts spreading; it becomes tough to control. Therefore, it is necessary to make the exact diagnosis of the disease and take remedial actions to prevent them in time. With the help of ML, early diagnosis of diseases is possible. Figure 7.11 illustrates the different diseases in the various crops such as, white, rust, fungal, blight, etc. Various ML and DL models used in the early detection of diseases and also the ways to handle the disease conditions.

(a) White Disease   (b) Rust Disease   (c) Fungal Disease   (d) Blight Disease

Figure 7.11 Various diseases in the crops and plants [8]. *Source:* kazakovmaksim/Adobe Stock Photos; Tunatura/Adobe Stock Photos; Yuliya/Adobe Stock; Skymet Weather Services Pvt. Ltd; darkfoxelixir/Adobe Stock Photos.

In general, to control plant disease, farmers spray the pesticides uniformly over the cropping area using measurable amounts of pesticides resulting in high financial cost. ML and DL algorithms give the optimal solution by taking agro-chemicals as input and time, place, and affected plants as target variables. With the help of ML and DL algorithms, early detection of plant disease is possible. Recognition of the disease at the earliest helps the farmers take the actions prior and stop the spread of disease. Thus, it enhances the crop yield and becomes financially strong. DL models predicted early threat or disease in vineyards by making farmers and winemakers aware of taking proper actions such as spread the pesticides in the field before-hand by processing hyperspectral agricultural data to extract correct information related to the disease [20]. Next, the SVM, AdaBoost, and RF ML models predicted the different diseases of grape plants (such as Black Rot, Esca, and Leaf Blight) by partitioning the diseased part into various segments. The diseased region is further divided into pieces using the threshold value to determine the correct disease [21].

Moreover, the SVM kernel functions and Gabor Wavelet Transform (GWT) detected and identified the type of disease, whether powdery mildew or early blight, in tomato leaf. Further, the K-means clustering algorithm extracted the input features, and SVM identified the disease in the pomegranate leaf. Using various explanatory features, the developed smartphone system, with Linear Support Vector Classification (SVC), K-NN, Extra Trees (ET) ML models, decided whether the disease is present or absent in Cassava plant. The diseases in Cassava plant are Cassava

Green Nite, Cassava Bacterial Blight, Cassava Brown Steak Disease, and Cassava Mosaic Disease. Next, the algorithms determined the severity of the infection also [22]. ML models, such as Gaussian Process Regression (GPR), Support Vector Regression (SVR), and Partial Least Square Regression (PLSR), determined the leaf rust disease of wheat plants using RGB data of disease-infected leaves. The CNN-based DL model predicted the disease in Paddy leaf using the image as input and image processing used to process the image. Next, CNN learns from the processed image and finds the leaf disease. To decide the infection in the plants, a camera sensor used to take image of the plant. Next, the image preprocessed with the help of image processing techniques. Further, the K-means algorithm divided the image into segments. Finally, the ANN Back Propagation-based ML model predicted the stem and leaf diseases such as late scorch, tiny whiteness, cottony mold, early scorch, ashen mold for the plants using the preprocessed image as input.

7.7 Future Research Directions

The vast and actual agricultural sensory data can introduce different technologies, such as transfer learning, incremental learning, Bayesian neural network, Internet of Things (IoT), genetic algorithms, and Fog computing architecture for making agricultural decisions. The farmers can see the field's status anytime and anywhere without considering any constraint on time and distance using IoT. IoT has its application in agriculture, known as the Internet of Agriculture Things (IoAT). The IoAT provides cost-effective and easy operations for in-situ soil and plant health information, early detection of climate change, etc. In general, the agricultural area is outside the village at remote places. So, reliable Internet connectivity is not available all the time. Therefore, edge computing can provide the solution by extracting meaningful data from extensive sensory data, reducing data size, and making it suitable for low Internet access. Long-range wireless communication protocols are beneficial for broad coverage and connectivity. Various agricultural fields contain a different set of sensors, creating multiple datasets. It is impractical to train numerous datasets to decide agrarian conditions. In this scenario, transfer learning offers a better choice. Game theory-based resource (such as carrier frequency, bandwidth, transmission power, etc.) allocation techniques can give a better performance in solving agricultural problems.

7.8 Conclusion

This chapter presented an overview of the CIA and usability of ML and DL techniques for modern agriculture. Also, we have illustrated and described various remote sensing methods, ML and DL models utilized for the agricultural field. Further, different wireless communication protocols available for agriculture are also covered. Finally, we have explained in detail the uses of various ML and DL models for estimation of accurate yield and improvement of crop yield, how to do water conservation and handle environmental change, how to improve soil and plant health, assessment of precise fertilizers and pesticides, and early detection of plant diseases.

Bibliography

1 Definition of agriculture. (2021). [Online]. Available: https://en.wikipedia.org/wiki/Agriculture, accessed April 6, 2021.

2 Image source for agricultural sensors. (2021). [Online]. Available: https://www.google.com/search?q=agricultural+sensors, accessed April 6, 2021.

3 Kumar, R., Mishra, R., Gupta, H.P., and Dutta, T. (2021). Smart sensing for agriculture: applications, advancements, and challenges. *IEEE Consumer Electronics Magazine, vol. 10, no. 4, pp. 51–56, July 1, 2021, doi: 10.1109/MCE.2021.3049623.*

4 Liakos, K.G., Busato, P., Moshou, D., Pearson, S., and Bochtis, D. (2018). Machine learning in agriculture: a review. *Sensors* 18 (8): 2674.

5 Chlingaryan, A., Sukkarieh, S., and Whelan, B. (2018). Machine learning approaches for crop yield prediction and nitrogen status estimation in precision agriculture: a review. *Computers and Electronics in Agriculture* 151: 61–69.

6 Kamilaris, A. and Prenafeta-Boldu´, F. X. (2018). Deep learning in agriculture: a survey. *Computers and Electronics in Agriculture* 147: 70–90.

7 W´ojtowicz, M., W´ojtowicz, A., Piekarczyk, J. et al. (2016). Application of remote sensing methods in agriculture. *Communications in Biometry and Crop Science* 11 (1): 31–50.

8 Image source for agricultural images. (2021). [Online]. Available: https://www.google.com/search?q=agriculture, accessed April 6, 2021.

9 Stas, M., Van Orshoven, J., Dong, Q., Heremans, S., and Zhang, B., (2016). A comparison of machine learning algorithms for regional wheat yield prediction using ndvi time series of spot-vgt. *2016 Fifth International Conference on Agro-Geoinformatics (Agro-Geoinformatics)*, 1–5. IEEE.

10 Heremans, S., Dong, Q., Zhang, B., Bydekerke, L., and Van Orshoven, J. (2015). Potential of ensemble tree methods for early-season prediction of winter wheat yield from short time series of remotely sensed normalized difference vegetation index and in situ meteorological data. *Journal of Applied Remote Sensing* 9 (1): 097095.

11 Nassar, L., Okwuchi, I.E., Saad, M., Karray, F., Ponnambalam, K., and Agrawal, P. (2020). Prediction of strawberry yield and farm price utilizing deep learning. *2020 International Joint Conference on Neural Networks (IJCNN)*, 1–7. IEEE.

12 Pantazi, X.E., Moshou, D., Alexandridis, T., Whetton, R.L., and Mouazen, A.M. (2016). Wheat yield prediction using machine learning and advanced sensing techniques. *Computers and Electronics in Agriculture* 121: 57–65.

13 Feng, Y., Peng, Y., Cui, N., Gong, D., and Zhang, K. (2017). Modeling reference evapotranspiration using extreme learning machine and generalized regression neural network only with temperature data. *Computers and Electronics in Agriculture* 136: 71–78.

14 Mohammadi, K., Shamshirband, S., Motamedi, S., Petkovi´c, D., Hashim, R., and Gocic, M. (2015). Extreme learning machine based prediction of daily dew point temperature. *Computers and Electronics in Agriculture* 117: 214–225.

15 Nahvi, B., Habibi, J., Mohammadi, K., Shamshirband, S., and Al Razgan, O.S. (2016). Using self-adaptive evolutionary algorithm to improve the performance of an extreme learning machine for estimating soil temperature. *Computers and Electronics in Agriculture* 124: 150–160.

16 Liang, L., Di, L., Zhang, L., Deng, M., Qin, Z., Zhao, S., and Lin, H. (2015). Estimation of crop lai using hyperspectral vegetation indices and a hybrid inversion method. *Remote Sensing of Environment* 165: 123–134.

17 Morellos, A., Pantazi, X.-E., Moshou, D., Alexandridis, T., Whetton, R., Tziotzios, G., Wiebensohn, J., Bill, R., and Mouazen, A.M. (2016). Machine learning based prediction of soil total nitrogen, organic carbon and moisture content by using vis-nir spectroscopy. *Biosystems Engineering* 152: 104–116.

18 Wang, J., Shen, C., Liu, N., Jin, X., Fan, X., Dong, C., and Xu, Y. (2017). Non-destructive evaluation of the leaf nitrogen concentration by in-field visible/near-infrared spectroscopy in pear orchards. *Sensors* 17 (3): 538.

19 Guo, P.-T., Li, M.-F., Luo, W., Tang, Q.-F., Liu, Z.-W., and Lin, Z.-M. (2015). Digital mapping of soil organic matter for rubber plantation at regional scale: an application of random forest plus residuals kriging approach. *Geoderma* 237: 49–59.

20 Hruška, J., Adão, T., Pádua, L., Marques, P., Peres, E., Sousa, A., Morais, R., and Sousa, J.J. (2018). Deep learning-based methodological approach for vineyard early disease detection using hyperspectral data. *IGARSS 2018–2018 IEEE International Geoscience and Remote Sensing Symposium*, 9063–9066. IEEE.

21 Patil, S.S. and Thorat, S.A. (2016). Early detection of grapes diseases using machine learning and IoT. *2016 Second International Conference on Cognitive Computing and Information Processing (CCIP)*, 1–5. IEEE.

22 Mwebaze, E. and Owomugisha, G. (2016). Machine learning for plant disease incidence and severity measurements from leaf images. *2016 15th IEEE International Conference on Machine Learning and Applications (ICMLA)*, 158–163. IEEE.

8

Long-and-Short-Term Memory (LSTM) Networks

Architectures and Applications in Stock Price Prediction

Jaydip Sen and Sidra Mehtab

Praxis Business School, Kolkata, India

8.1 Introduction

The recurrent neural networks (RNNs) are a special type of neural network that is capable of processing and modeling sequential data such as text, speech, time-series, etc. In RNNs, the output of the network at a given time slot depends on the current input to the network as well as the previous state of the network [1]. Unfortunately, these networks are poor in capturing the long-term dependencies in the data due to a problem known as the vanishing or exploding gradients [2]. LSTM networks, a variant of RNNs, have the ability to overcome the problem of vanishing or exploding gradients, and hence such networks are quite effective and efficient in analyzing time series and other sequential data. LSTM networks consist of gates that are essentially memory cells in a computing machine. The gates store past information about their states and control the information flow over time. There are four types of gates in LSTM networks. The *forget gates* decide what information from the past to discard and what to retain at a given state. The input gates enable the network to control the input at the current state. The contents of the forget gates and the input gates are aggregated into a cell state vector. In other words, the cell state aggregates the old state information from the forget gate with the current information from the input gate. Finally, the output gate yields the output from the networks at the current time slot. The unique design of LSTM networks and the use of the *backpropagation through time* (BPTT) algorithm enable these networks to effectively process sequential data like text, speech, and time series.

This chapter first presents the working principles of RNNs in processing sequential data and discusses the adverse effect of the exploding and vanishing gradients in such networks. As a solution to such a problem, the chapter introduces LSTM networks, their architectural design, and operations. Several variants of LSTM architecture are also discussed. Finally, the application of LSTM models in stock price prediction is presented in detail.

The chapter is organized as follows. Section 2 presents a discussion on the architecture of RNNs and the vanishing and exploding gradient problem that these networks usually suffer from. Section 3 presents the general architecture of an LSTM node and discusses the functionalities of different gates. Section 4 identifies stock price prediction as one of the applications of LSTM networks and presents a brief

Emerging Computing Paradigms: Principles, Advances and Applications, First Edition.
Edited by Umang Singh, San Murugesan and Ashish Seth.
© 2022 John Wiley & Sons Ltd. Published 2022 by John Wiley & Sons Ltd.

overview of some related work. Section 5 presents the design of six different architectures of LSTM models with a particular focus on the stock price prediction. Section 6 provides some experimental results on the performances of the models in predicting future stock prices. Finally, Section 7 concludes the chapter.

8.2 RNN Architecture Vanishing and Exploding Gradient Problems

The RNNs recurrently apply the same function over a sequence. The operations in an RNN can be expressed by the recurrence Equation (8.1).

$$S_t = f\left(S_{t-1}, X_t\right) \tag{8.1}$$

In (8.1), S_{t-1} denotes the state of the network at the time slot t-1, X_t is the input to the network at the time slot t, S_t is the state of the network at the time slot t, and f is the function used to compute S_t. The equation expresses how the states of the network are computed sequentially over different time slots using its previous state information and the input at the current state.

In (8.1), the function f is a differentiable function also known as the activation function. For most practical applications, f is chosen as the *hyperbolic tangent* function. Hence, (8.1) can be written alternatively as in (8.2).

$$S_t = tanh\left(S_{t-1} * W + X_t * U\right) \tag{8.2}$$

In (8.2), W represents a linear transformation that maps one state to another state, and U is a linear mapping function from the input set to the set of the states. The other popular activation functions are *sigmoid*, *rectified linear unit* (ReLU), other variants of ReLU [3–6]. In Figure 8.1, O_t is the output of the network at time slot t, and it is given by (8.3).

$$O_t = V * S_t \tag{8.3}$$

In an RNN, state information at each time slot is computed using all previous state information using (8.1). This implies that an RNN should remember information over an arbitrarily long period of time. However, in practice, state information over only a

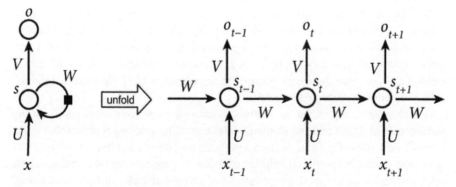

Figure 8.1 Schematic representation of RNN architecture and its states at three consecutive time slots *t−1, t* and *t+1*.

limited number of time slots in the past is used by RNNs to counter the *vanishing and exploding gradient problem*. This will be discussed briefly later in this section.

RNNs are trained using an algorithm known as BPTT, which is a modified version of the well-known *backpropagation* algorithm [7–9]. The difference between BPTT and backpropagation is needed to handle the extra complexity arising out of the multiple inputs in RNNs consisting of the previous states' information S_{t-1}, the current input X_t, and the parameters W and U, which are shared between each layer in the network.

The working principle of the classical backpropagation algorithm in neural networks is based on propagating the error gradient back from the output layer to the first hidden layer. The gradient of the *cost function* (i.e., the error function) with respect to each parameter in the network (i.e., the weights of the links and the biases at the nodes) are first computed. Using these gradients, the algorithm updates each parameter based on a gradient descent step. However, the execution of the algorithm quite often leads to increasingly smaller gradients for the lower layers. The smaller gradient values lead to a situation in which the successive updates of the weights for the lower layers by the gradient descent become negligible. This leads to a suboptimal convergence for the training of the network. This problem is known as the vanishing gradient problem [10–12]. In some situations, the opposite problem may occur, wherein the magnitudes of the gradients increase out of bound, leading to very large updates of the weights for the lower layers. The backpropagation algorithm does not converge in this case too, and the network suffers from the *exploding gradient problem*. The vanishing and exploding gradient problems in RNNs can also be illustrated as follows.

As the first forward pass of the backpropagation algorithm finishes its execution, the derivative of the error function is computed with respect to each parameter (i.e., the weights of the links and the biases of the nodes) of the network. Once this gradient is computed, it is propagated back through the stack of activities that were done in the forward step. This backward pass pops activities off the stack to accumulate the error derivatives at each time slot. The recurrence relation exhibiting the gradient propagation through the network is represented in (8.4).

$$\frac{\partial E}{\partial S_{t-1}} = \frac{\partial E}{\partial S_t}\frac{\partial S_t}{\partial S_{t-1}} = \frac{\partial E}{\partial S_t}W \tag{8.4}$$

From (8.4), using the state change over m time slot, it is evident that the recurrence relation that propagates the gradient backward through time in an RNN forms a geometric sequence as presented in (8.5).

$$\frac{\partial S_t}{\partial S_{t-m}} = \frac{\frac{\partial S_t}{\partial S_{t-1}}*\ldots\ldots*\frac{\partial S_{t-m+1}}{\partial S_{t-m}}}{\partial S_{t-m}} = W^m \tag{8.5}$$

It is clear from (8.5) that the gradient grows exponentially if $|W| > 1$, causing the exploding gradient problem. On the other hand, the gradient shrinks exponentially if $|W| < 1$, leading to the vanishing gradient problem. If the weight parameter W is a matrix instead of a scalar, then this gradient problem is related to the largest eigenvalue (ρ) of W. The largest eigenvalue is also known as the spectral radius. If $\rho < 1$, the gradients progressively vanish. The exploding gradient problem arises when $\rho > 1$ [13].

The exponential growth or decay of gradients in deep neural networks can be tackled following three approaches. The three approaches are mentioned below:

- *Truncated backpropagation through time (TBPTT)*: this method is also known as the *gradient clipping* method, in which the gradients above a given threshold value are clipped [11, 14, 15]. While TBPTT effectively counters the exploding gradient problem, it suffers from a drawback. The clipping of the gradient puts a constraint on the number of rounds on the backpropagation leading to non-optimal values of the parameters.
- Use of optimization methods such as momentum, RMSProp, AdaGrad, and Adam that do not rely much on the local gradients [16–20].
- LSTM networks, a variant of RNN architecture for which the training converges much faster while effectively capturing the long-term dependencies exhibited by the data [4, 21–25].

In the sections that follow, we discuss the architecture of LSTM networks and their application in time series analysis with a particular focus on the stock price prediction.

8.3 General Architecture of LSTM Nodes

The concept of the LSTM network was first introduced in the literature to overcome the *vanishing and decreasing gradient problems* of RNNs [22]. The fundamental block of an LSTM network is a memory cell. In essence, a memory cell is a hidden layer in a neural network. Every memory cell has a recurrent edge. The values carried by the recurrent edge are known as the states of the cell. The recurrent edge should ideally be associated with a link weight $w = 1$ to ensure that the vanishing and exploding gradient problems do not occur. An exploded form of an LSTM cell is presented in Figure 8.2.

The state information at the previous time slot, $C^{(t-1)}$, is transformed into the cell state at the current time slot $C^{(t)}$ via a number of intermediate operations instead of

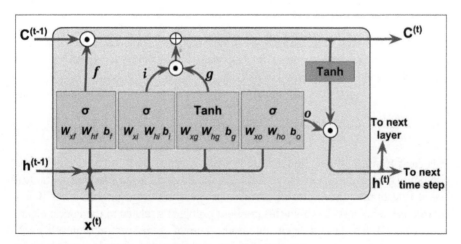

Figure 8.2 The structure of an LSTM node [Adapted from [2]].

being directly multiplied by a weight factor [2]. The information flow in the memory cell of an LSTM node is controlled by some gates, which will be discussed in what follows. In Figure 8.2, the operation ⊙ denotes element-wise multiplication, and the operator ⊕ stands for element-wise addition. At the time slot t, the input to the LSTM node is denoted as $x^{(t)}$. The hidden unit at time slot t -1 is represented as $h^{(t-1)}$.

The four boxes marked with yellow color depict four gates in the LSTM node. The inputs to the boxes are aggregated, and an activation function (either the *sigmoid* or the *hyperbolic tangent*) is applied to the aggregated input to produce the outputs from the boxes. For computing the aggregate input in a box, vector multiplications of the inputs with the weight vectors are performed.

A typical LSTM node consists of four different types of gates—the forget gate, the input gate, the input modulation gate, and the output gate. The functionalities of these gates are discussed below.

1) *Forget gate*: This gate enables the node to reset the state of the node by discarding past information, which is not of much use currently. It effectively decides which information from the previous state is to be allowed to get into the node and which information is to be discarded. The output of the forget gate at time slot t, f_t is given by (8.6):

$$f_t = \sigma\left(W_{xf} * x^{(t)} + W_{hf} * h^{(t-1)} + b_f\right) \tag{8.6}$$

In (8.1), σ represents the sigmoid activation function, W_{xf} denotes the weight matrix for the input and the forget gate, W_{hf} is the weight matrix for the previous hidden layer and the forget gate, and the b_f is the bias of the forget gate node, $x^{(t)}$ is the input at time slot t.

2) *Input gate and Input modulation gate*: The *input gate* and the *input modulation gate* update the cell state of the LSTM node. The outputs of the input gate (i_t) and the input modulation gate (g_t) are given by (8.7) and (8.8), respectively.

$$i_t = \sigma\left(W_{xi} * x^{(t)} + W_{hi} * h^{(t-1)} + b_i\right) \tag{8.7}$$

$$g_t = tanh(W_{xg} * x^{(t)} + W_{hg} * h^{(t-1)} + b_g) \tag{8.8}$$

Using the values of i_t and g_t as derived in (8.7) and (8.8), respectively, the state of the node at time slot t is computed using (8.9).

$$C^{(t)} = (C^{(t-1)} \odot f_t \oplus \left(i_t \odot g_t\right) \tag{8.9}$$

3) *Output gate*: The output yielded by the output gate at time slot t is computed using (8.10).

$$o_t = \sigma\left(W_{xo} * x^t + W_{ho} * h^{(t-1)} + b_o\right) \tag{8.10}$$

Finally, using the values of o_t in (8.10), the hidden units at time slot t are computed using (8.11).

$$h^{(t)} = o_t \odot tanh(C^{(t)}) \tag{8.11}$$

There are many variants of LSTM proposed in the literature other than the one whose architecture we have described [26]. Gated Recurrent Unit (GRU) is a very popular variant that is computationally more efficient due to its simpler architecture than LSTMs [27].

8.4   Stock Price Prediction–An Application of LSTM

As mentioned earlier in the chapter, LSTM networks are particularly well-suited for processing sequential data, e.g., time-series data, text, speech, and video. In this chapter, we will discuss the application of LSTM in time series analysis and forecasting, with particular on the stock price prediction. In this section, we present a brief discussion on some of the existing approaches to stock price prediction.

The literature on stock price prediction models and portfolio optimization is quite rich. Several approaches have been proposed by researchers for accurate prediction of future values of stock prices and using the forecasted results in building robust and optimized portfolios that optimize the returns while minimizing the associated risk. Time series decomposition and econometric approaches like *autoregressive integrated moving average* (ARIMA), Granger causality, *vector autoregression* (VAR) are some of the most popular approaches to future stock price predictions that are used for robust portfolio design [28–33]. The use of machine learning (ML), deep learning (DL), and reinforcement learning models for future stock price prediction has been the most popular approach of late [34–42]. Hybrid models utilize the algorithms and architectures of ML and DL and exploit the sentiments in the textual sources on the social web [43–47]. The use of *generalized autoregressive conditional heteroscedasticity* (GARCH) in estimating the future volatility of stocks and portfolios is a very popular approach [51].

8.5   Different LSTM Architectures for Stock Price Prediction

In this section, we present six different variants of LSTM architectures for stock price prediction. The six LSTM models are built based on these architectures. The models are trained and tested using the historical index values of NIFTY 50. NIFTY 50 index is the weighted average of the index values of the top 50 stocks listed on the National Stock Exchange of India [48]. The multivariate model LSTM Model #4 uses all the five features of the stock price data, viz., *open*, *high*, *low*, *close*, and *volume*. The other five models use univariate input *open*. Except for the LSTM Model #1 that has an input of the past five days' daily *open* values, all other models are based on the input of the past two weeks' data. All models predict the next five days' *open* values. Assuming five working days for a stock exchange, the models predict the *open* value for the next five days.

8.5.1   LSTM Model #1

The last week's *open* values of the stock are used as the input to this model. Hence, the input data is univariate in nature. On the basis of the univariate input data, the *open* values of the next five days are predicted by the model. Hence, the shape of the input is (5, 1). The input gets into an LSTM block containing 200 nodes. A brute force search method using the grid-search method is used to determine the number of nodes in the LSTM block that optimizes the execution time and the accuracy of the model. A fully connected block with 100 nodes receives the output of the LSTM block. The output from the fully connected layer (i.e., the dense layer) is received by the final output layer, which contains five nodes. These five nodes yield the next five predicted values in sequence. An epoch value of 20 is used for training the model, while the size of each batch is 16. The design of the LSTM Model #1 is depicted in Figure 8.3.

8.5.2 LSTM Model #2

This LSTM model uses as its input the *open* values of the stock of the last two weeks. Based on the input data, the model predicts the *open* values of the stock for the five days in the next week. Except for this difference in the input data, all other design parameters and hyperparameters of this model are similar to those used in the LSTM Model #1. The input to the model has a shape (10, 1). Here, 10 represents the size of the input (i.e., the number of days of which the *open* values are used as the input), and 1 represents the feature count (here the *open* values). The design of the model is exhibited in Figure 8.4.

8.5.3 LSTM Model #3

The design of this model is uniquely characterized by the inclusion of an encoder and decoder blocks for encoding and decoding the features extracted from the time series of historical stock prices. The input to the model is the *open* values of the stock of the

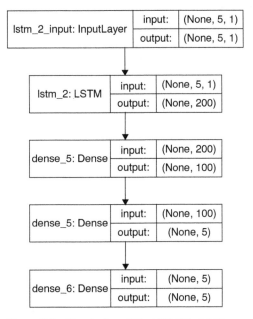

Figure 8.3 The design of the LSTM Model #1.

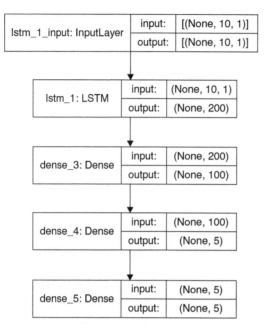

lstm_1_input: InputLayer	input:	[(None, 10, 1)]
	output:	[(None, 10, 1)]

lstm_1: LSTM	input:	(None, 10, 1)
	output:	(None, 200)

dense_3: Dense	input:	(None, 200)
	output:	(None, 100)

dense_4: Dense	input:	(None, 100)
	output:	(None, 5)

dense_5: Dense	input:	(None, 5)
	output:	(None, 5)

Figure 8.4 The design of the LSTM Model #2.

last ten days. The model includes two LSTM blocks. While the first LSTM block encodes the features from the input data, the second block is responsible for decoding the encoded features. There are 200 nodes in the encoder LSTM block. Since the model uses the last ten days' *open* values at its input, the input data shape for the first LSTM block (i.e., the encoder LSTM) is (10, 1). Each of the nodes in the encoder block extracts one feature from the ten *open* input values. The 200 features extracted by the encoder LSTM block are stored in a one-dimensional vector of size 200. As the output of the encoder block is received by the repeat vector block, for each timestamp in the output sequence of the model, the repeat vector block extracts the input data features once. It may be noted here that the predicted values from the model have five timestamps in its sequence corresponding to the five consecutive predicted *open* values. The *repeat vector* layer yields an output data shape of (5, 200). The data shape indicates that for each timestamp (there are five timestamps), 200 features from the input data are used to sequentially produce the output.

The decoder LSTM block's output is sent to a dense layer. The learning from the decoded features happens in the dense layer. The predicted *open* values for the next five days are produced through the output nodes of the dense layer. However, unlike the LSTM Model #1 and the LSTM Model #2, the predicted values of this model are not available at the same timestamp. Instead, these values are computed and produced by the model in five separate rounds. The TimeDistributedWrapper function defined in Keras is responsible for synchronizing the operations of the decoder LSTM block, the dense layer, and the final output layer. This ensures the sequential operations of each round are time-synchronized, and the output at each round does not intermingle with the output of the previous or the next round. The number of epochs used in training the model is 70 epochs, while each batch has a size of 16. The schematic design of the LSTM Model #3 model is exhibited in Figure 8.5.

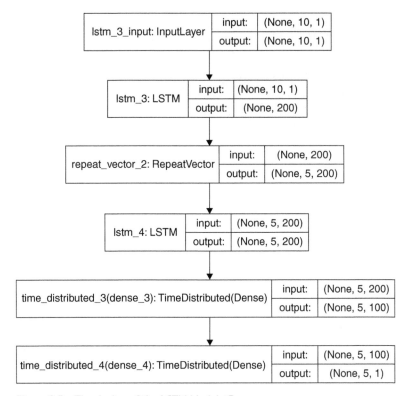

Figure 8.5 The design of the LSTM Model #3.

8.5.4 LSTM Model #4

The LSTM Model #4 is a multivariate variant of the LSTM Model #3. The input to the model is the last couple of weeks' stock price data containing all the five features. The input data have the shape (10, 5). The training of the model is done using a batch size of 16 and 20 epochs. The design of the LSTM Model #4 is presented in Figure 8.6.

8.5.5 LSTM Model #5

This is an adapted variant of the LSTM Model #3. The encoding of the input data is carried out by a convolutional neural network (CNN) sub-model [1]. CNNs are not particularly well-suited for learning from sequential data like time series. However, one-dimensional CNNs are extremely powerful in performing feature extraction from time-series data. In this model, features from the input data are extracted by a CNN sub-module, which are then provided as the input to the LSTM module for decoding. The decoding of the features is performed by the LSTM layer, and at the end, the future values are predicted sequentially. The CNN layer is designed to include two convolution blocks, each having 64 feature maps and kernels of size 3. Since the input to the model is the *open* values of the last couple of weeks, the input data shape is (10, 1). The first convolutional layer

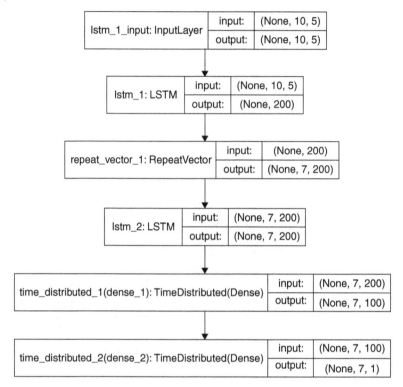

Figure 8.6 The design of the LSTM Model #4.

produces an output data shape (8, 64). The value of 8 is derived as (10-3+1). The value 64 indicates the dimension of the feature space.

The second convolutional layer produces an output data shape (6, 64). The dimension of the feature space is contracted by 1/2 by a max-pooling layer that follows the second convolutional layer. Accordingly, the max-pooling layer yields an output data shape (3, 64). Note that the second parameter in the data shape is not changed by a pooling layer. The max-pooling layer's output is converted into a one-dimensional vector by a *flatten* operation. The size of this one-dimensional array is 3*64 = 192. The decoding LSTM layer containing 200 nodes receives this one-dimensional array of size 192 as the input. The design of the decoder LSTM layer of this model is exactly similar to that of the LSTM Model #3. For the training and the testing of the model, 20 epochs and a batch size of 16 are used. The design of the LSTM Model #5 is shown in Figure 8.7.

8.5.6 LSTM Model #6

This model is an adapted variant of the LSTM Model #5. The convolution operations of the encoder CNN and the decoding operations of the LSTM layer are combined for each output sequence round. This hybrid model, known as the Convolutional-LSTM model, efficiently combines the convolutional operation

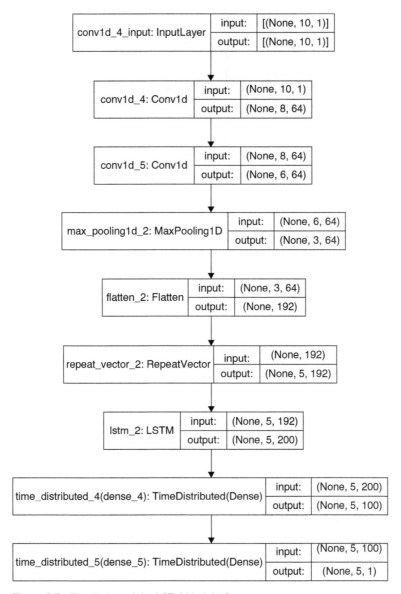

Figure 8.7 The design of the LSTM Model #5.

with the LSTM decoding [49, 50]. This model reads the input sequence of the time series, carries out the convolution operation for encoding, and performs the decoding of the extracted features at the LSTM layer. The ConvLSTM2d class of Keras framework contains a method that is capable of executing convolution operations in two dimensions with an integrated LSTM block for decoding [49]. The two-dimensional ConvLSTM class can be adapted for processing univariate data with one input feature. Figure 8.8 depicts the schematic design of the LSTM Model #6.

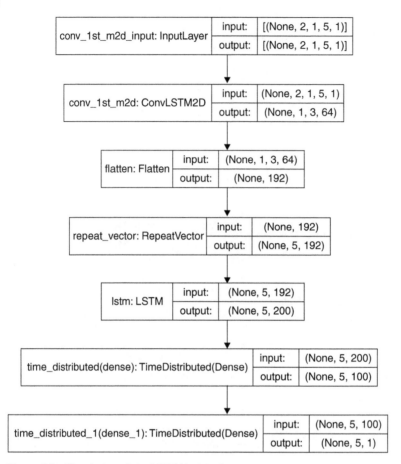

Figure 8.8 The design of the LSTM Model #6.

8.6 Experimental Results

The six LSTM models discussed in Section 5 are executed on a computing machine consisting of an i7-9750H CPU with a clock speed of 2.60–2.59 GHz. Python 3.7.4, TensorFlow 2.3.0, Keras 2.4.3, and various other associated modules are used in implementing the models. The historical daily NIFTY 50 index values from December 29, 2014 to December 28, 2018 are used for training the models. The models are tested on the NIFTY 50 index from December 31, 2018 to July 31, 2020. The training data comprise 1,045 records, while 415 records are used for testing. Each model is run for ten rounds, and the mean RMSE of the ten rounds is computed. The mean execution times for the ten rounds are also computed. The mean *open* value of the test dataset is 11,070.59. The ratio of the RMSE of each model to the mean *open* value is also computed. Table 8.1 presents the comparative performance of the models in terms of their prediction accuracies. The LSTM Model #1 is found to be the most accurate one as its RMSE, and hence RMSE/Mean value is the lowest. The multivariate LSTM Model #4 is the least accurate one yielding the highest value of RMSE. It is observed that the basic LSTM models (LSTM #1 and #2) are generally more accurate than

Table 8.1 Accuracies of the LSTM models.

Model	RMSE	RMSE/Mean	Rank
LSTM Model #1	344.57	0.0311	1
LSTM Model #2	390.46	0.0353	3
LSTM Model #3	408.97	0.0369	4
LSTM Model #4	1,893.85	0.1711	6
LSTM Model #5	460.11	0.0416	5
LSTM Model #6	388.77	0.0350	2

Table 8.2 Execution speeds of the LSTM models.

Model	Exec. Time (sec)	Rank
LSTM Model #1	18.64	4
LSTM Model #2	31.44	5
LSTM Model #3	14.53	2
LSTM Model #4	66.91	6
LSTM Model #5	15.25	3
LSTM Model #6	11.17	1

their encoder-decoder counterparts. While the multivariate LSTM model is comparatively less, all six models are highly precise. The highest value of the ratio of the RMSE to the mean *open* values is 0.1711. This implies that the least accurate model has committed a 17.11% percent error over the mean value of the target.

Table 8.2 presents the average execution speeds of the six models. The mean execution time of the LSTM Model #6 for ten rounds is found to be the lowest. The multivariate LSTM Model #4 is found to be the slowest one requiring an average time of 66.91 seconds for a single round. It is interesting to note that the encoder-decoder univariate models (LSTM #3, #5, and #6) are faster than the general LSTM univariate models (LSTM #1 and #2), and the multivariate LSTM model (LSTM #4).

8.7 Conclusion

In the feed-forward neural networks, data can travel only from the input layer to the output layer via the intermediate hidden layers. These networks are not capable of processing and analyzing sequential data like text, speech, video, and time-sequential observations. RNNs with links connecting the higher layers to the lower layers enable data communication in the backward direction as well. These networks are quite effective in processing sequential data. However, RNNs suffer from the vanishing and exploding gradient problem that leads to suboptimal training, and hence, inaccurate models. A variant of RNNs known as LSTM networks effectively gets rid of the vanishing and exploding gradient problem.

LSTMs have been proved to be very efficient and accurate in handling sequential data. This chapter first introduced the basic design of LSTM networks and highlighted their working principles. Subsequently, six different variants of LSTM models were presented. The models differ in their input data shape, constituent building blocks, and the way the outputs are produced. As an application of LSTM networks, the six models are then used for stock price prediction. Using NIFTY 50 index values from December 29, 2014 to December 28, 2018, the models are trained. The models are then tested on NIFTY 50 records from December 31, 2018 to July 31, 2020. For evaluating the performance, the models are executed for ten rounds, and the average performance over the ten rounds is considered as the overall performance. The evaluation of the models is done on two metrics: (i) the average RMSE to the mean *open* value and (ii) the average execution time. The results revealed some very interesting observations. From an accuracy perspective, the univariate LSTM model with the last week's *open* values as the input was found to be most accurate. Further, the basic LSTM models with univariate inputs were found to be more precise than the encoder-decoder variants. The encode-decoder convolutional LSTM was found to be the fastest in execution, while the multivariate encoder-decoder variant was the slowest. The encoder-decoder univariate LSTM models were found to be faster than the basic LSTM models. Optimization of the model further using different regularization approaches, different activations functions, and optimizers is a possible future direction of research.

Bibliography

1 Geron, A. (2019). *Hands-On Machine Learning with Scikit-Learn Keras & Tensorflow*, 2e. Sebastopol, CA: O'Reilly Media Inc. ISBN-13: 978-1492032649.

2 Sebastian, R. and Mirjalili, V. (2019). *Python Machine Learning*, 3e. Birmingham, UK: Packt Publishers. ISBN-13: 978-1789955750.

3 Bing, X., Wang, N., Chen, T., and Li, M. (2015). Empirical evaluation of rectified activations in convolution network. *arXiv:1505.oo853v2*, 27 November 2015.

4 Dieng, A. B., Ranganath, R., Altosaar, J., and Blei, D.M. (2018). Noisin: unbiased regularization for recurrent neural networks. *Proceedings of the 35th International Conference on Machine Learning (ICML'18)*, Stockholm, Sweden (10–15 July).

5 Rasamoelina, A. D., Adjailia, F., and Sincak, P. (2020). A review of activation function for artificial neural network. *Proceedings of 2020 IEEE 18th World Symposium on Applied Machine Intelligence and Informatics (SAMI)*, Herlany, Slovakia, 281–286 (January 23–25). doi: 10.1109/SAMI48414.2020.9108717.

6 Seo, J., Lee, J., and Kim, K. (2017). Activation functions of deep neural networks for polar decoding applications. *Proceedings of the 2017 IEEE 28th Annual International Symposium on Personal, Indoor, and Mobile Radio Communications (PIMRC)*, Montreal, QC, Canada, 1–5 (October 8–13). doi: 10.1109/PIMRC.2017.8292678.

7 Bersini, H. and Gorrini, V. (1997). a simplification of the backpropagation-through-time algorithm for optimal neurocontrol. *IEEE Transactions on Neural Networks* 8 (2): 437–441. doi: 10.1109/72.557698.

8 Lillicrap, T. P. and Santoro, A. (2019). Backpropagation through time and the brain. *Current Opinion in Neurobiology* 55: 82–89. doi: 10.1016/j.conb.2019.01.011.

9 Werbos, P. J. (1990). Backpropagation through time: what it does and how to do it. *Proceedings of the IEEE* 78 (10): 1550–1560. doi: 10.1109/5.58337.

10 Bengio, Y., Simard, P., and Frasconi, P. (1994). Learning long-term dependencies with gradient descent is difficult. *IEEE Transactions on Neural Networks* 5 (2): 157–166. doi: 10.1109/72.279181.

11 Pascanu, R., Mikolov, T., and Bengio, Y. (2012). Understanding the exploding gradient problem. arXi:1211.5063. https://arxiv.org/abs/1211.5063 (accessed July 22, 2021).

12 Rebeiro, A., Tiels, K., Aguirre, L. A., and Schon, T. B. (2020). Beyond exploding and vanishing gradients: analysing RNN training using attractors and smoothness. *Proceedings of the 23rd International Conference on Artificial Intelligence and Statistics (AISTATS'20)*, Palermo, Italy. http://proceedings.mlr.press/v108/ribeiro20a/ribeiro20a.pdf (accessed July 22, 2021)

13 Pascanu, R., Mikolov, T., and Bengio, Y. (2013). On difficulty of training recurrent neural networks. *Proceedings of the 30th International Conference on Machine Learning (ICML'13)*, June, Atlanta, Georgia, USA, 28, 1310–1318.

14 Puskorius, G. V. and Feldkamp, L. A. (1994). Truncated backpropagation through time and Kalman filter training for neurocontrol. *Proceedings of the 1994 IEEE International Conference on Neural Networks (ICNN'94)*, Orlando, FL, USA, 2488–2493 (June 28–July 2, 1994). doi: 10.1109/ICNN.1994.374611.

15 Seetharaman, P., Wichern, G., Pardo, B., and Roux, J. L. (2020). Autoclip: adaptive gradient clipping for source separation networks. *Proceedings of the 2020 IEEE 30th International Workshop on Machine Learning for Signal Processing (MLSP'20)*, Espoo, Finland, 1–6 (September 21–24). doi: 10.1109/MLSP49062.2020.9231926.

16 Duchi, J., Hazan, E., and Singer, Y. (2011). Adaptive subgradient methods for online learning and stochastic optimization. *Journal of Machine Learning Research* 12: 2121–2159.

17 Hinton, G., Srivastava, N., and Swersky, K. (2014). Neural networks for machine learning. https://www.cs.toronto.edu/~hinton/coursera/lecture6/lec6.pdf (accessed July 22, 2021).

18 Kingma, D. and Ba, J. (2015). Adam: a method for stochastic optimization. *Proceedings of the 3rd International Conference on Learning Representations (ICLR'15)*, San Diego, CA, USA (May 7–9).

19 Sutskever, I., Martens, J., Dahl, G., and Hinton, G. (2013). On the importance of initialization and momentum in deep learning. *Proceedings of the 30th International Conference on Machine Learning (ICML'13)*, Atlanta, GA, USA, III, 1139–1147 (June 16–21).

20 Zhong, H., Chen, Z., Qin, C., Huang, Z., Zheng, V.W., Xu, T., and Chen, E. (2020). Adam revisited: a weighted past gradients perspective. *Frontiers of Computer Science* 14: 145309. doi: 10.1007/s11704-019-8457-x.

21 Graves, A. (2014). Generating sequences with recurrent neural networks. arXiv:1308.0850v5. https://arxiv.org/pdf/1308.0850.pdf (accessed July 22, 2021).

22 Hochreiter, S. and Schmidhuber, J. (1997). Long short-term memory. *Neural Computation* 9 (8): 1735–1780. doi: 10.1162/neco.1997.9.8.1735.

23 Liu, Y. and Howard, M. W. (2020). Generation of scale-invariant sequential activity in linear recurrent networks. *Neural Computation* 32 (7): 1379–1407. doi: 10.1162/neco_a_01288.

24 Sak, H., Senior, A., and Beaufays, F. (2014). Long short-term memory recurrent neural network architecture for large scale acoustic modeling. *Proceedings of INTERSPEECH'14*, Singapore, 338–342 (September 14–18).

25 Zaremba, W., Sutskever, I., and Vinyals, O. (2015). Recurrent neural network regularization. arXiv:1409.2329v5. https://arxiv.org/pdf/1409.2329.pdf (accessed July 22, 2021).

26 Jozefowicz, R., Zaremba, W., and Sutskever, I. (2015). An empirical exploration of recurrent network architecture. *Proceedings of the 32nd International Conference on Machine Learning (ICML'15)*, Lille, France, 15, 2342–2350.

27 Chung, J., Gulcehre, C., Cho, K., and Bengio, Y. (2014). Empirical evaluation of gated recurrent neural networks on sequence modeling. *Proceedings of NIPS Workshop on Deep Learning* (December 2014).

28 Sen, J. (2018). Stock composition of mutual funds and fund style: a time series decomposition approach towards testing for consistency. *International Journal of Business Forecasting and Marketing Intelligence (IJBFMI)* 4 (3): 235–292. doi: 10.1504/IJBFMI.2018.092781.

29 Sen, J. and Datta Chaudhuri, T. (2016a). An investigation of the structural characteristics of the Indian IT sector and the capital goods sector: an application of the R programming language in time series decomposition and forecasting. *Journal of Insurance and Financial Management* 1 (4): 68–132. doi: 10.36227/techrxiv.16640227.v1.

30 Sen, J. and Datta Chaudhuri, T. (2016b). An alternative framework for time series decomposition and forecasting and its relevance for portfolio choice: a comparative study of the Indian consumer durable and small cap sectors. *Journal of Economics Library* 3 (2): 303–326. doi: 10.1453/jel.v3i2.787.

31 Sen, J. and Datta Chaudhuri, T. (2017a). A predictive analysis of the Indian FMCG sector using time series decomposition-based approach. *Journal of Economics Library* 4 (2): 206–226. doi: 10.1453/jel.v4i2.1282.

32 Sen, J. and Datta Chaudhuri, T. (2017b). A time series analysis-based forecasting framework for the Indian healthcare sector. *Journal of Insurance and Financial Management* 3 (1): 66–94. doi: 10.36227/techrxiv.16640221.v1.

33 Sen, J. and Datta Chaudhuri, T. (2018). Understanding the sectors of the Indian economy for portfolio choice. *International Journal of Business Forecasting and Marketing Intelligence (IJBFMI)* 4 (2): 178–222. doi: 10.1504/IJBFMI.2018.090914.

34 Bao, W., Yue, J., and Rao, Y. (2017). A deep learning framework for financial time series using stacked autoencoders and long-and-short-term memory. *PLoS ONE* 12 (7): e0180944. doi: 10.1371/journal.pone.0180944.

35 Binkowski, M., Marti, G., and Donnat, P. (2018). Autoregressive convolutional neural networks for asynchronous time series. *Proceedings of the 35th International Conference on Machine Learning (ICML'18)*, Stockholm, Sweden, 580–589 (July 10–15).

36 Mehtab, S. and Sen, J. (2020a). Stock price prediction using CNN and LSTM-based deep learning models. *Proceedings of the IEEE International Conference on Decision Aid Sciences and Applications (DASA)*, Sakheer, Bahrain, 447–453 (November 8–9, 2020). doi: 10.1109/DASA51403.2020.9317207.

37 Mehtab, S. and Sen, J. (2020b). Stock price prediction using convolutional neural network on a multivariate time series. *Proceedings of the 3rd National Conference on Machine Learning and Artificial Intelligence (NCMLAI'20)*, New Delhi, India (February 1). doi: 10.36227/techrxiv.15088734.v1.

38 Mehtab, S. and Sen, J. (2021). A time series analysis-based stock price prediction using machine learning and deep learning models. *International Journal of Business Forecasting and Marketing Intelligence (IJBFMI)* 6 (4): 272–335. doi: 10.1504/IJBFMI.2020.115691.

39 Mehtab, S., Sen, J., and Dasgupta, S. (2020a). Robust analysis of stock price time series using CNN and LSTM-based deep learning models. *Proceedings of the IEEE 4th International Conference on Electronics, Communication and Aerospace Technology (ICECA'20)*, Coimbatore, India, 1481–1486 (November 5–7, 2020). doi: 10.1109/ICECA49313.2020.9297652.

40 Mehtab, S., Sen, J., and Dutta, A. (2020b). Stock price prediction using machine learning and LSTM-based deep learning models. In: *Machine Learning and Metaheuristics Algorithms and Applications (Somma'20)* (ed. S.M. Thampi et al.), Vol. 1386, 88–106. Singapore: Springer. doi: 10.1007/978-981-16-0419-5_8.

41 Sen, J., Dutta, A., and Mehtab, S. (2021a). Profitability analysis in stock investment using an LSTM-based deep learning model. *Proceedings of the IEEE 2nd International Conference on Emerging Technologies (INCET'21)*, Belgaum, India, 1–9 (May 21–23). doi: 10.1109/INCET51464.2021.9456385.

42 Sen, J. and Mehtab, S. (2021). Accurate stock price forecasting using robust and optimized deep learning models. *Proceedings of the IEEE International Conference on Intelligent Technologies (CONIT'21)*, Hubli, India (June 25–27, 2021). doi: 10.1109/CONIT51480.2021.9498565.

43 Bollen, J., Mao, H., and Zeng, X. (2011). Twitter mood predicts the stock market. *Journal of Computational Science* 2 (1): 1–8. doi: 10.1016/j.jocs.2010.12.007.

44 Chen, M.-Y., Liao, C.-H., and Hsieh, R.-P. (2019). Modeling public mood and emotion: stock market trend prediction with anticipatory computing approach. *Computers in Human Behavior* 101: 402–408. doi: 10.1016/j.chb.2019.03.021.

45 Galvez, R.H. and Gravano, A. (2017). Assessing the usefulness of online message board mining in automatic stock prediction systems. *Journal of Computational Science* 19: 43–56. doi: 10.1016/j.jocs.2017.01.001.

46 Mehtab, S. and Sen, J. (2019). A robust predictive model for stock price prediction using deep learning and natural language processing. *Proceedings of the 7th International Conference on Business Analytics and Intelligence (BAICONF'19)*, Bangalore, India (December 5-7). doi: 10.36227/techrxiv.15023361.v1.

47 Weng, B., Ahmed, M. A., and Megahed, F. M. (2017). Stock market one-day ahead movement prediction using disparate data sources. *Expert Systems with Applications* 79: 153–163. doi: 10.1016/j.eswa.2017.02.041.

48 NSE. (2021). NIFTY 50 Index. https://www.nseindia.com/products-services/indices-nifty50-index (accessed July 22, 2021).

49 Shi, X., Chen, Z., Wang, H., Yeung, D.-Y., and Wong, W.-K. (2015). Convolutional LSTM network: a machine learning approach for precipitation nowcasting. *Proceedings of the 28th International Conference on Neural Information Processing Systems*, Cambridge, MA, USA, 1, pp. 802–881 (December 7–12).

50 Weidman, S. (2019). *Deep Learning Form Scratch: Building with Python from First Principles*, 1e. USA: O'Reilly Media Inc.

51 Sen, J., Mehtab, S., and Dutta, A. (2021b). Volatility modeling of stocks from selected sectors of the Indian economy using GARCH. *Proceedings of the IEEE Asian Conference on Innovation in Technology (ASIANCON'21)* (28-29 August 2021).

Additional Reading

Ding, B., Qian, H., and Zhou, J. (2018). Activation functions and their characteristics in deep neural networks. *Proceedings of 2018 Chinese Control and Decision Conference (CCDC)*, Shenyang, China, 1836–1841 (9–11 June). doi: 10.1109/CCDC.2018.8407425.

Ledoit, O. and Wolf, M. (2008). Robust performance hypothesis testing with the Sharpe ratio. *Journal of Empirical Finance* 15 (5): 850–859. doi: 10.1016/j.jempfin.2008.03.002.

Sen, J. (2018). Stock price prediction using machine learning and deep learning frameworks. *Proceedings of the 6th International Conference on Business Analytics and Intelligence (ICBAI'18)*, Bangalore, India (20–21 December 2018).

Sen, J. and Mehtab, S. (2021). Design and analysis of robust deep learning models for stock price prediction. In: *Machine Learning: Algorithms, Models and Applications* (ed. J Sen). London, UK: IntechOpen Publishers. doi: 10.5772/intechopen.99982.

Sen, J. and Datta Chaudhuri, T. (2016). Decomposition of time series data of stock markets and its implications for prediction: an application for the Indian auto sector. *Proceedings of the 2nd National Conference on Advances in Business Research and Management Practices (ABRMP'16)*, Kolkata, India, 15–28 (8–9 January 2016). doi: 10.13140/RG.2.1.3232.0241.

Part 4

Advances in Wireless Networks

9

Mobile Networks

5G and Beyond

Pavel Loskot

ZJU-UIUC Institute, Zhejiang, Haining, 718 East Haizhou Road, China

9.1 Cellular Networks and Mobility

The emergence of mobile cellular networks in the late 80s revolutionized the delivery of public telecommunication services. For the first time, the phone calls could be made anywhere and anytime from portable battery-powered devices within the coverage of cellular base stations. The evolution of mobile cellular networks since then followed the path of digitalization, increasing the data rates, improving the coverage and reliability, and offering a plethora of new services and applications to the end users. The design of cellular networks has been initially optimized only for spectral efficiency. Later energy efficiency became important to prolong the battery life, and especially to sustain profits for the network operators as their operational and capital expenditures grew significantly over time. The recent 5G networks offer ultra-reliable and low latency communications for users, machines, and devices. The application requirements in wireless networks must reflect the trade-offs between data rates, link reliability and latency, and the achievable coverage.

In general, the mobility in cellular networks must be supported across all protocol layers. At the physical layer, the key task is to establish and maintain connectivity among the neighboring nodes. The traveling user must undergo periodic hand overs between neighboring base stations in order to maintain an uninterrupted connectivity to the core network. At upper layers, the user mobility is supported as long as the network can be constantly aware of the user's present location. It requires that both the users and all the base stations are assigned globally unique identifiers or addresses in order to redirect application traffic to the desired location within the network. The key principles of mobility support in the current generations of cellular networks at the physical and upper layers are discussed in the following two subsections.

9.1.1 Mobility Management at Physical Layer

Figure 9.1 depicts a typical hand over scenario when the user equipment (UE) is moving between two neighboring base stations (BS). As the UE is moving away from one base station closer to its cell edge, the signal strength of the BS and the

Emerging Computing Paradigms: Principles, Advances and Applications, First Edition.
Edited by Umang Singh, San Murugesan and Ashish Seth.

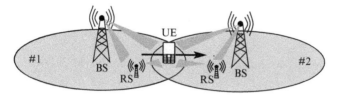

Figure 9.1 Coordinated transmissions between base stations and the cell hand over.

achievable data rates rapidly decrease due to a fundamental exponential propagation law. At the same time, the signal strength from the following BS starts to increase, which may create a co-channel interference and further deteriorate the link quality. The 4G networks (known as the Long-Term Evolution (LTE)) address this issue by employing advanced multiple antenna techniques.

- Coordinated beamforming and scheduling: The transmissions among multiple base stations are coordinated to reduce or even completely null their interference, and to increase the data throughput and the coverage. This strategy is often referred to as coordinated multi-point (CoMP).
- Improved localization of the UE: The beamforming can be assisted by information about the UE location. The localization in the 4G network mainly uses the enhanced cell identifiers (extended with signal strength and angle-of-arrival measurements), and the observed time differences of arrival.
- Relaying: The relays can act as simplified base stations, and create small in-band or out-band overlay cells. The priority is to remove the coverage holes. It is important to balance the traffic load among the main and overlay cells.
- Self-Organizing Networks (SONs): This functionality is a key enabler for distributed cell overlays. An automated optimization of the cell coverage and its capacity also assists the user mobility. For example, the hand over parameters are automatically adjusted to avoid unnecessary and failed handovers, to manage the interference, and to improve the stability of radio links.
- Distributed sub-carrier allocation: The mobile users can mitigate the frequency selectivity of wideband propagation channels by distributing their transmissions over non-contiguous blocks of sub-carriers.

The hand over (also referred to as hand-off) procedures are driven by periodic measurements of the link quality and the related performance metrics. The measurement reports are evaluated to decide whether the hand over criteria are met. The actual hand over involves exchanging signaling messages and reallocating the network resources. The 3GPP LTE standard states that mobility across the cellular network should be supported for speeds up to 350 km/h in the cells with 5 km radius, and a modest degradation in the network service quality should be tolerable in the cells with up to 30 km radius.

9.1.2 Mobility Management at Upper Layers

Since the inception of mobile cellular networks, the upper layer mobility management has evolved around maintaining the permanent as well as

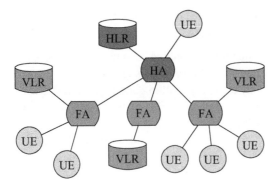

Figure 9.2 Mobility management of the UE using Mobile IP and the distributed registry.

temporary user profile data including the current cell, geographical location and the service subscription status. The UE is registered and de-registered with the network as it is turned on or powered down, respectively. The UE data are distributed across the network in the control centers. The unique international mobile subscriber identity (IMSI) is often used as the database key. The databases are referred to as home and visitor location register (HLR and VLR), respectively. The location updates are periodically performed by the UE to maintain up-to-date location information in the registry. The update procedures must be carefully designed to minimize unnecessary transmissions of messages and to reduce other mobility management costs. A common approach is to trigger the registry updates by various events such as a hand over or a key parameter change, and by expiration timers.

The most common mobility management protocols in cellular mobile networks are Mobile Internet Protocol (IP) and Session Initiation Protocol (SIP). The latter protocol is intended for managing multimedia traffic whereas Mobile IP exploits distributed databases maintaining up to date information about the UE. The HLR and VLR registry in Mobile IP are interconnected via home and foreign agents (HA and FA) as shown in Figure 9.2. Upon registration in the visited cell with a FA, the current UE location is updated both in VLR and HLR by exchanging messages between the FA and the HA. It ensures that the current UE location can be queried at any time from its HLR, and subsequently a route to the UE determined. However, this mechanism becomes rather inefficient when the UE is highly mobile, so frequent updates of its location in HLR would be required. One solution to this problem is to enable exchange of messages between the old FA and the new FA.

9.2 Mobile Networks

Mobile ad-hoc networks (MANETs) can be deployed and operated nearly anywhere, since they are not constrained by the availability of supporting infrastructure. However, this deployment flexibility may entail severe penalties in the performance, complexity, scalability, stability, and reliability of the offered network services.

Depending on the scenario, a MANET can consist of up to hundreds of nodes. The design of MANETs faces several crucial challenges.

- The trajectories of MANET nodes must be coordinated to at least some degree in order to avoid collisions and to constrain the internode distances.
- The internode distances, and possibly also propagation conditions, and thus, the node connectivity are time varying. Consequently, the routes in MANETs have limited life span, and so must be periodically re-established.
- The energy is required not only to provide communication services, but also to propel the nodes. This significantly reduces the operational life span of nodes and of the whole network in comparison to the traditional handheld devices. Consequently, it is often beneficial to balance the traffic load and the energy consumption equally among the network nodes.
- The node trajectories can be optimized to fulfill a given task with a minimum time or energy. This often leads to difficult optimization problems.
- There is a trade-off in achieving the maximum coverage and maintaining the network connectivity. The latter is important for mission critical applications.
- MANET applications are typically provisioned in a fully distributed manner in order to create robustness against node failures, and also to balance the use of transmission and computing resources.

Transmission protocols in MANETs must handle not only mobility and dynamic network topology, but also provide sufficient level of network autonomy with self-organizing and self-healing capabilities. Traditional networking protocols such as TCP/IP are less suitable for use in dynamic wireless networks such as MANETs due to large protocol overheads. For instance, dynamic assignment of IP addresses in highly mobile MANETs is likely to be too slow. It is much more efficient to use globally unique identifiers instead of IP addresses in MANETs in order to support the decentralization and scalability.

The transmission overhead is caused by frequent retransmissions over unreliable links, and by the need to encrypt the transmitted messages. The overhead can be reduced using header compression schemes, but in practice, it is often preferred to develop proprietary protocols. The proprietary protocols are more secure, and unlike TCP/IP protocols, can be highly optimized for the given MANET deployment scenario. The disadvantage of using proprietary protocols is possible compatibility issues in addition to substantial efforts and time required to develop the protocols that are reliable and thoroughly tested.

The slotted multiple-access (MAC) protocols require time and frequency synchronization of the network nodes. The precise clock synchronization in wireless networks can be obtained by distributing a single master clock throughout the network using Precision Time Protocol (PTP). Time synchronization using global satellite navigation signals may not be accurate enough, and such signals may not be always available e.g., due to obstacles or jamming.

The network topology in MANETs is formed in three steps: neighbors discovery, route discovery, and route maintenance. The neighbor discovery can be done by broadcasting hello messages. In order to manage the routing complexity and improve the scalability, a two-level hierarchical topology is usually assumed. The nodes are

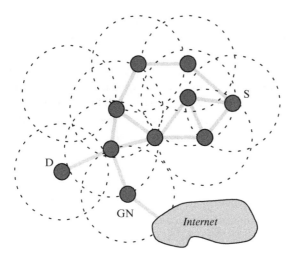

Figure 9.3 A MANET topology with one gateway node (GN) to the Internet and multiple routes between the source node S and the destination node D.

clustered, and the cluster heads are elected by a suitable procedure. The cluster heads should be periodically re-elected in order to share the energy consumption among nodes fairly. The internode communications only occur via the cluster heads. Furthermore, one or more cluster heads can act as a gateway to the outside Internet as shown in Figure 9.3.

Since the nodes are normally connected to their respective cluster heads via a single hop, the multi-hop routing is concerned with connecting the cluster heads. The routing algorithm can minimize packet loss, round-trip time (RTT) and other costs such as the number of hops and the energy consumption. Some links can be asymmetric with different costs in opposite directions. More importantly, the cumulative costs of routes in the network can be constrained in order to satisfy the application requirements. For instance, in mission critical applications, there can be a deadline by which the message must reach its destination. In order to make the routes more stable and reliable, a common strategy is to determine more than one route between the source-destination pairs which are critical to the application or the network operation. The additional routes serve as a failback option, or they can be used to aggregate the transport capacity. It is also important that all routes are loop-free, and that the overhead required for the route setup and maintenance is minimized.

There are two basic approaches to routing in MANETs. Proactive routing utilizes routing tables. These tables must be regularly updated, which may incur a large overhead. This routing offers a fast response to the routing requests, can maintain backup routes, and it is well-suited for high traffic in hierarchical networks with low mobility. The other strategy is reactive or on-demand routing. This routing has the advantage that the routes are established only when needed, but the associated overhead can be also a problem. The previously determined routes can be cached to facilitate a faster route discovery. Although reactive routing has slower response than proactive routing, it has better scalability, and is more suitable for networks with flat topology, high mobility nodes, and lower traffic demands.

The protocol performance can be improved by exploiting location information [1]. The mobility requires that the node locations are periodically estimated, although their trajectories can be also predicted. Multiple nodes and the mobility itself provide the opportunity to obtain enough measurements to estimate the locations of all nodes in the network. The challenge is the measurement accuracy, for example, due to non-line of sight propagation, unintentional or intentional interference, and the clock synchronization errors.

The design of MANETs requires to assume realistic models of radio propagation and of node mobility. For propagation models, it is often sufficient to assume a detailed link budget including the path-loss and random shadowing in order to accurately estimate the received signal strength. For some MANET scenarios such as tactical missions, it is common to use commercial software tools to plan beforehand the connectivity of MANETs in the difficult terrain of deployment. The accurate mobility models must account for spatiotemporal correlations of node trajectories, and other factors including mobility decisions to various events, and mobility adaptation to a terrain profile. In addition, some MANETs such as tactical mobile networks must assume multiple radio propagation and mobility models for their design and deployment.

9.2.1 Computing Applications in MANETs

The mobility poses a challenge not only to locate the client and the server, but also to maintain the state of the application session. The key requirement is that the in-network client and server are independent of the current network topology. The cluster heads can act as local storage or cache for application data in addition to routing application data between the server and the current location of the client. Suspending and subsequently resuming the session is less practical in dynamic MANETs where the applications are often context aware. For instance, the application should migrate and be executed only in the nodes residing in a given geographical area, or only on the nodes with enough available resources. The application migration can be implemented, for example, using virtual machines, or using smart messages. The latter approach combines data and instructions how to process the data when they traverse the network. This provides a very flexible mechanism to control the data processing and their delivery. The data instructions account for the node context including its location, available resources and other attributes such as the time bounds for data delivery. The messages can be aggregated, or new messages generated. It is also possible to define the message admission and execution control rules.

Off-loading computations to the cloud may have unacceptable latency. The solution can be partial off-loading and maintaining a copy of the application state in the mobile node and on the server. The application calls can be then executed on the server or by the node client to optimize the use of resources.

Distributed MANET applications are often implemented as a middleware layer that is inserted in between the operating system and the application code. The middleware can ensure consistency of data and computations, maintain a synchronized global state across the network, manage remote procedure calls, and provide abstraction and logical addressing for the resources and other entities residing in the layers below. The middleware enables to flexibly add and remove the resources and nodes from the distributed application as required. The middleware functionality also enables to greatly simplify the orchestration of otherwise complex distributed network resources and services.

The real-time (RT) applications can provide soft (on average, or with some probability) or hard (maximum or the worst case) guarantees on the latency of the application response. The latency guarantees can be achieved by reservation of the network resources, e.g., considering multiple routes between sources (i.e., event reporting nodes) and destinations (i.e., decision nodes). The node types can vary in time. If the latency limits are exceeded, a failure is declared. In MANETs, the application response latency is primarily given by the RTT of message exchange. The RT applications rely on prioritization of tasks and processes with pre-emptive interrupts, and priority scheduling of messages. The challenge is that the tasks performed in MANET nodes are a combination of concurrent and sequential processes with random execution times. For randomly occurring events, the question is how often the application needs to check for these events to satisfy the latency constraint. The RT considerations can be generalized to quality-of-service (QoS) provisioning.

Figure 9.4 shows a distributed architecture of computing application in MANETs. The data processing flows are represented by a directed acyclic graph (DAG). The decision-making node initiates the computing via a control node. The control node distributes the instructions to other nodes using smart messages. This architecture implements a map-compute-reduce-decide model. It is well-suited for detecting and evaluating ambient events in the environment.

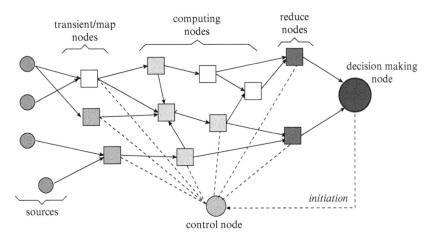

Figure 9.4 The distributed architecture of computing application in a MANET.

9.3 Mobile Networks in the 5G

The recent 5G cellular networks largely adopted physical layer solutions from the preceding 4G networks. The greatest challenge in designing 5G networks is orchestration of a very large pool of transmission and computing resources distributed across different types of networks. The objective is a seamless provisioning of diverse applications and services for a very large number of mobile users and devices as they move across different wireless access networks outdoors as well as indoors. The complexity of managing the 5G networks dictates the use of autonomous sub-systems, self-configuring, self-optimizing and self-organizing networks, and automated monitoring with predictive and prescriptive data processing methods. The 5G networks offer new licensed and unlicensed spectrum sharing models. All these approaches inherently create super-flexibility where software updates are now more important for the network performance than hardware updates [2]. It allows to pool the distributed network resources as required, and to exploit different levels of abstraction and virtualization for unified presentation of resources to applications.

Unlike the previous generations of mobile cellular networks, the architecture of 5G networks allows integration of whole MANETs in addition to traditional support of individual mobile users and devices as depicted in Figure 9.5. The VANETs supported in the 5G involve self-driving connected cars, drones (UAVs), high-altitude platforms (HAPs), and the low-Earth orbit (LEO) satellite networks [3]. These networks allow to create complex cyber-physical systems including self-driving vehicles, smart cities, smart factories, and other such systems. The VANETs enhance the capability of the underlying physical system (e.g., improve safety and efficiency), but they can also serve as an extension of the 5G infrastructure to improve the communication coverage and capacity.

All types of VANETs supported by the 5G networks must consider the following design aspects that are often interrelated.

- Network size and mobility of nodes: The mobility affects the internode distances, and thus, the signal attenuation. The mobility must be coordinated to avoid physical collisions while maintaining the node connectivity.
- Radio-wave propagation: The network nodes usually remain at line-of-sight visibility. However, nearby obstacles can create non-line-sight propagation conditions in connectivity to the supporting infrastructure.

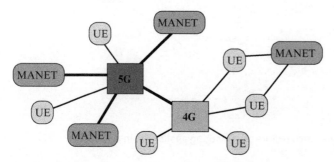

Figure 9.5 The 5G extended support for MANETs in addition to traditional UE.

- Connectivity among nodes and to the supporting infrastructure: The connectivity can involve radio-frequency (RF) as well as free-space optical (FSO) transmissions. The challenge is how to design antennas to align the beam patterns of transmitting and receiving antennas. This is a three-dimensional rather than two-dimensional problem while the antennas are mounted on highly mobile nodes.
- Interference management: VANETs are overlay networks above the terrestrial cellular infrastructure. This may create new interference patterns.

9.3.1 Connected Cars

Connecting the vehicles is the first necessary step towards self-driving vehicles and their full autonomy. The initial inter-vehicle communication systems were created and standardized as dedicated short-range communications (DSRC). These networks enabled direct device-to-device links between nearby vehicles and to the dedicated roadside infrastructure. The peer-to-peer and broadcast messages could be sent with priority, low latency and secure, guaranteed delivery. The primary driver was to improve traffic safety. However, the emergence of the 4G and now 5G networks made the DSRC system obsoleted, so its spectrum was recently released and reallocated to the unlicensed spectrum.

The current motivations for connecting the cars are rather broad. The driver assistance system offers safety related applications including collision avoidance, intersection crossing, parking, and lane changing assistance. Traffic management applications provide navigation services, optimize car routes and reduce traffic congestion. Traffic information systems are used for digital maps, toll payments, and police checks. There are also applications for in-car entertainment and general Internet access. More advanced applications enabling different levels of self-driving require the use of sensory systems such as Lidar and Radar to understand and predict the car surrounding environment.

The rich connectivity is also being created inside cars as their digitalization progresses. This generates large volumes of data that can be monetized, provided that the privacy issues are also considered. The data together with machine learning algorithms enable new applications such as predicting failures and optimizing the car operations. More importantly, the generated data are processed inside the car, and only the results of this processing are transmitted to the external network, saving significant amount of bandwidth.

9.3.2 Unmanned Aerial Vehicles (UAVs)

Traffic rules for road vehicles are well defined, however, the airspace management for commercial and civilian use of UAVs is much less understood and developed. The UAVs pose new and sometimes severe risks to safety and security. For instance, even a lightweight UAV may cause severe injury or death of a person, if it falls uncontrollably to the ground. Small and cheap UAVs carrying explosives represent a significant security concern as they can damage expensive infrastructure. Effective protection of the infrastructure against the drone attacks is an urgent technical problem to solve. The lightweight UAVs are susceptible to weather conditions, and their operation is problematic near obstacles and outside the direct visibility of the ground operator.

Developing a collision avoidance system for UAVs is rather problematic, since their flexible flight dynamics makes their trajectories much less predictable. The UAVs cannot solely rely on satellite navigation systems, since these systems may not be available around high rise obstacles such as buildings and mountains.

In order to ensure a compliance with the government regulations on the save and legal use of UAVs, the certified UAV operators and providers may need to be established, for example, to enable drone-as-a-service (DaaS). The main civilian use of UAVs includes communication applications and infrastructure monitoring. The UAVs can carry simplified base stations and provide immediate line-of-sight coverage in the recovery and emergency communication networks. The UAVs are used to provide effective monitoring of buildings, agricultural crops, and transmission lines and pipes in the remote areas. The UAVs are used for surveillance and tracking, search and rescue missions, locating forest fires, and data collection from wireless sensors. The payload delivery by drones is attractive for providing the healthcare support in remote areas.

9.3.3 Tactical Mobile Networks

Tactical mobile networks interconnect large number of networks and equipment mounted on different types of vehicles with possibly very different mobility. These networks are organized in hierarchical tiers starting from low-level portable radios up to the use of geostationary (GEO) communication and navigation satellites enabling global roaming. Tactical networks provide fully integrated communication environment to carry out missions with information flows between the battlefield and the central command. The requirements for tactical networks are more demanding than for their civilian counterparts. Tactical network must show jamming resilience and provide information assurance in different harsh environments. The in-built security is achieved by be spoke proprietary communication protocols. This may create issues with interoperability of telecommunication equipment supplied by different vendors.

Tactical communication equipment is typically designed as software defined radio (SDR) to enable a multitude of radio interfaces and protocols. The basic communication services in tactical networks include delivery of command & control (C2) messages, full motion video, and the encrypted information relying and backhaul. The promise and ubiquitous coverage of 5G networks has not gone unnoticed by the developers of tactical mobile networks. There is a significant ongoing interest in adopting commercial off-the-shelf communication technology for use in tactical networks. The main motivation is to reduce the development efforts and costs and shorten the time to market. However, the adoption is challenging as commercial telecommunication equipment does not meet many of the security and reliability standards expected in tactical networks. Moreover, defense markets have different dynamics, and the frequent updates of technology is considered to be undesirable.

9.3.4 LEO Satellites

The low-Earth orbit (LEO) satellites are deployed to altitudes between 200 and 2,000 km above the Earth surface. The LEO satellites approximately follow the Kepler orbits (ignoring e.g., the atmospheric drag) with typical orbital periods of about 2 hours. The LEO communication satellites are often lightweight and cheap. Moreover, the recently established companies (e.g., Space X) enabled cheap launches to the orbit. The ultimate objective is to deploy a sufficient number of LEO satellites to achieve a ubiquitous Earth coverage to provide affordable Internet access for a small monthly subscription fee.

There are, however, some serious technological challenges that may not be possible or at least not easy to overcome.

- Weak received signals: In comparison to terrestrial cellular networks, the distances to satellites are substantially larger. However, the number and size of antennas used on handheld devices and on the satellites is limited.
- Non-homogeneous user density: The density of terrestrial users varies substantially from little to very large. However, the small size satellites may only support a modest number of simultaneous connections.
- Intermittent coverage: The full Earth coverage by LEO satellites will require large scale satellite constellations and their careful planning. However, the larger the satellite network, the more difficult it is to prevent collisions, manage the interference, and avoid traffic congestion and excessive latencies.

The above issues can be overcome by establishing dedicated terrestrial satellite access stations outside the densely populated areas in order to primarily support communication services for users in the remote areas. Assuming a relatively small number of terrestrial stations offering large bandwidth to the satellites will have significant impact on the design of LEO satellite networks.

9.4 Long-Term Perspectives

Providing communication services with good quality of experience to mobile users remains one of the key drivers in evolving mobile cellular networks [4]. It has been argued that the current 5G networks offer sufficient flexibility and advances in exploiting the existing telecommunication equipment that the future generations of cellular networks will mainly concern software updates. The defense industry showed great interest to incorporate 5G equipment to tactical mobile networks that will likely affect the development of 5G networks.

The wireless connectivity has matured. Quantum communications may bring the desired breakthrough to physical connectivity, although this will likely come from physics researchers than from telecommunication engineers. There are still opportunities for developing energy efficient hardware including integrated circuits, and for designing antennas suitable for mobility with many degrees of freedom.

As the networks are becoming more complex, there is a growing concern how to configure, manage, and monitor all the available communication and computing resources in order to maximize their usage efficiency and to provide seamless services to the users. The complexity and the reliance on communication services tacitly creates serious security issues. Therefore, robustness and resilience are becoming as important as the efficiency in designing mobile cellular networks. For mission critical cyber-physical systems, communication networks must be designed as dependable real-time systems.

The main issue with the UAV networks is the airspace management and providing the safety guarantees. It remains to be seen how the LEO satellite networks are going to be utilized, but there is certainly a need to provide connectivity for users in remote areas and to travelers onboard long-haul vehicles.

9.5 Conclusion

The chapter outlined key principles governing the design of mobile networks that are now directly integrated with the 5G cellular networks. The mobile networks provide communication and computing services to the underlying cyber-physical systems. The key and often conflicting communication drivers in developing mobile networks are flexibility, efficiency, autonomy, robustness, and security. The most notable mobile networks include connected cars, UAV and LEO satellite networks, and more encompassing tactical mobile networks. These networks share similar requirements and challenges, and their operation is often dependent on the availability of satellite navigation systems.

Bibliography

1 Akyidliz, I.F., Kak, A., and Nie, S. (July 2020). 6G and beyond: the future of wireless communications systems. *IEEE Access* 8: 133995–134030. doi: 10.1109/ ACCESS.2020.3010896.

2 Al-Shehri, S.M., Loskot, P., and Hirsch, M.J. (2019). Localization enhanced mobile networks in mobile computing. In: (ed. J.H. Ortiz). IntechOpenSci.

3 Benhaddou, D. and Al-Fuqaha, A. (2015). *Wireless Sensor and Mobile Ad-Hoc Networks*. Springer.

4 Chowdhury, M.Z., Shahjalal, M.D., Ahmed, S., and Jang, Y.M. (August 2020). 6G wireless communication systems: applications, requirements, technologies, challenges, and research directions. *IEEE Access* 1: 957–975. doi: 10.1109/ OJCOMS.2020.3010270.

5 Dahlman, E., Parkvall, S., and Sköld, J. (2014). *4G: LTE/LTE-Advanced for Mobile Broadband*, 2e. Elsevier.

6 Dang, S., Amin, O., Shihada, B., and Alouini, M.-S. (January 2020). What should 6G be? *Nature Electronics* 3: 20–29. doi: 10.1038/s41928-019-0355-6.

7 Elnashar, A., El-saidny, M.A., and Sherif, M.R. (2014). *Design, Deployment and Performance of 4G-LTE Networks*. Wiley.

8 EU H2020 Project No. 763601. DroC2om: drone critical communications. https:// www.droc2om.eu, 2017–2019.

9 Fitzek, F. and Seeling, P. (March 2020). Why we should NOT talk about 6G. arXiv:2003.02079[cs.NI].

10 Giordani, M., Polese, M., Mezzavilla, M., Rangan, S., and Zorzi, M. (March 2020). Toward 6G networks: use cases and technologies. *IEEE Communications Magazine* 58 (3): 55–61. doi: 10.1109/MCOM.001.1900411.

11 Gui, G., Liu, M., Tang, F., Kato, N., and Adachi, F. (October 2020). 6G: opening new horizons for integration of comfort, security, and intelligence. *IEEE Wireless Communications* 27 (5): 126–132. doi: 10.1109/MWC.001.1900516.

12 Guillen-Perez, A. and Cano, M.-D. (October 2018). Flying ad hoc networks: a new domain for network communications. *Sensors* 18 (3571): 1–23. doi: 10.3390/s18103571.

13 Jamalipour, A. and Ma, Y. (2011). *Intermittently Connected Mobile Ad Hoc Networks*. Springer.

14 Loskotetal, P. (February 2015). Long-term socio-economical drivers of traffic in next generation broadband networks. *Annals of Telecommunications* 70 (1–2): 1–10.

15 Lu, N., Cheng, N., Zhang, N., Shen, X., and Mark, J.W. (August 2014). Connected vehicles: solutions and challenges. *IEEE Internet of Things* 1 (4): 289–299. doi: 10.1109/JIOT.2014.2327587.

16 Morgan-Jones, I. and Loskot, P. (October 2019). Regional coverage analysis of LEO satellites with Kepler orbits. arXiv:1910.10704[physics.space-ph].

17 Mozaffari, M., Saad, W., Bennis, M., and Debbah, M. (January 2019). Communications and control for wireless drone-based antenna array. *IEEE Transactions on Communications* 67 (1): 820–834. doi: 10.1109/TCOMM.2018.2871453.

18 Mozaffari, M., Saad, W., Bennis, M., Nam, Y.-H., and Debbah, M. (March 2019). A tutorial on UAVs for wireless networks: applications, challenges, and open problems. *IEEE Communications Surveys & Tutorials* 21 (3): 2334–2360. doi: 10.1109/COMST.2019.2902862.

19 Nawaz, H., Ali, H.M., and Laghari, A.A. (March 2020). UAV communication networks issues: a review. *Archives of Computational Methods in Engineering* 1–21. doi: 10.1007/s11831-020-09418-0.

20 Roy, R.R. (2011). *Handbook of Mobile Ad Hoc Networks for Mobility Models*. Springer.

21 Sesia, S., Toufik, I., and Baker, M. (2009). *LTE – The UMTS Long Term Evolution*. Wiley.

22 Varrall, G. (2018). *5G and Satellite Spectrum, Standards, Scale*. Artech.

23 Wang, X. (2011). *Mobile Ad-hoc Networks: Protocol Design*. InTech Open.

24 Xiang, W., Zheng, K., and Shen, X. (2017). *5G Mobile Communications*. Springer.

25 Xu, W., Zhou, H., Cheng, N., Lyu, F., Shi, W., Chen, J., and Shen, X. (January 2018). Internet of vehicles in Big data era. *IEEE/CAA Journal of Automatica Sinica* 5 (1): 19–35. doi: 10.1109/JAS.2017.7510736.

26 Yang, D., Jiang, K., Zhao, D., Yu, C., Cao, Z., Xie, S., Xiao, Z., Jiao, X., Wang, S., and Zhang, K. (September 2018). Intelligent and connected vehicles: current status and future perspectives. *Science China Technological Sciences* 61: 1–26. doi: 10.1007/s11431-017-9338-1.

27 Yanmaz, E., Yahyanejad, S., Rinner, B., Hellwagner, H., and Bettstetter, C. (January 2018). Drone networks: communications, coordination, and sensing. *Ad Hoc Networks* 68: 1–15. doi: 10.1016/j.adhoc.2017.09.001.

28 Zhang, L., Zhao, H., Hou, S., Zhao, Z., Xu, H., Wu, X., Wu, Q., and Zhang, R. (July 2019). A survey on 5G millimeter wave communications for UAV-assisted wireless networks. *IEEE Access* 7: 117460–117504. doi: 10.1109/ACCESS.2019.2929241.

10

Advanced Wireless Sensor Networks

Research Directions

Richa Sharma

Assistant Professor, ITS Engineering College, Greater Noida

10.1 Advanced Wireless Sensor Networks

WSN have shown their potential in various application domains like military monitoring, health, industrial control, weather monitoring, commodity tracking, home control, cognitive sensing and spectrum management, security management, etc. Few of the recent advances in this technology are as sensor devices localization, designing smart home/smart office environments, supporting several military operations, advancement in several industrial or commercial processes, traffic monitoring, and management at expressways or highways healthcare monitoring, etc.

Smart environments at homes or offices provides control to the owner of their homes or of their work places by automating different systems like controlling lightning systems, controlling window blinds, electrical appliances in kitchens, or security system. WSN technology helps in managing combat operations efficiently. Sensing technology can be used to detect the movement of intruder units on land or sea side. With the tremendous rise in the technological advances in wireless technology, sensor nodes are also playing important roles in industrial applications in terms of efficient supervisory control, optimizing product's quality improvement, and data acquisition. These nodes are deployed in industry areas for temperature monitoring, measuring pressure parameters, checking the flow level, etc.

Traffic congestion is one of the biggest challenges in every big metropolitan city. This kind of severe traffic congestion can be alleviated by properly managing the traffic. Multiple sensor devices can be deployed in heavy traffic prone areas to automatically collect real-time traffic data. Collection must be employed for efficient management of rush-hour traffic. Several researches are going to design an Intelligent Transport Management Systems to surface transportation by making use of sensor technology. The tracking of vehicle movement on road also termed as vehicular monitoring to locate it is also one of the applications of sensor technology. Due to limited life of the power level, it becomes important to monitor these sensors timely. Structure monitoring is also done using sensors through their inspection at regular intervals of time. This kind of monitoring is done to repair these structures to maintain their better working conditions. This helps in reducing the maintenance cost

Emerging Computing Paradigms: Principles, Advances and Applications, First Edition.
Edited by Umang Singh, San Murugesan and Ashish Seth.
© 2022 John Wiley & Sons Ltd. Published 2022 by John Wiley & Sons Ltd.

and in preventing harmful incidents occurring due to building collapsing. Smart agriculture can also be achieved by introducing WSN concept to collect timely information related to soil degradation or scarcity of water resource. Healthcare monitoring through the sensing devices helps in collecting real-time data about health parameters of the patients.

In all these above-mentioned application areas, there is a need to focus on all components of energy dissipation in WSNs separately. Due to the high integration of these components within a WSN, and therefore their interplay, each component cannot be treated independently without regard for other components; in another words, optimizing the energy consumption of one component, e.g., MAC protocols, may increase the energy requirements of other components, such as routing. Therefore, minimizing energy in one component may not guarantee optimization of the overall energy usage of the network. This view of overall energy consumption in WSNs can be applied to optimizing and balancing energy consumption and increasing the network lifetime. Like most of the other network types like Adhoc networks, mobile ad-hoc networks (MANETs), vehicle ad-hoc networks (VANETs), research in the field of sensor networks is also surrounded by several other research sub domains. These sub domains are minimization of network congestion, ensuring high network coverage, providing time synchronization among nodes, TDMA scheduling, and secure communication among the nodes in the network.

Due to the random deployment of the sensors, redundancy exists in the services provided by these devices like sensing data about physical parameters or while routing data to other nodes. To make WSN fully operational or functional, several design issues are yet to be studied and resolved. For instance, sensor node's resources such as memory, communication bandwidth, and most importantly their energy need to be effectively managed to prolong their lifetime and productivity. Further, nodes should be assigned different roles carefully to prolong network lifetime. Since usually all these sensors are deployed in remote and unattended regions, proper deployment, and maintenance is very much essential. LEACH [1] is one of the distributed approaches that surpasses conventional static clustering or direct transmission by performing randomized rotation of cluster heads. LEACH has also been considered as a benchmark till date by several researchers for energy efficient working of WSNs.

During, past few years, these research fields have gained enormous attention of researchers and practitioners. All these research areas focus primarily on the fulfilment of a common objective that is balancing of node energy consumption. In addition, all these research fields in some respect also show dependency on each other. For example, an optimized node deployment is essential to ensure high network connectivity and complete network area coverage.

10.2 Research Dimensions of Advanced WSN

Several research dimensions of advanced WSN such as network congestion, node deployment, data aggregation, clustering, and base station positioning are discussed below along with the research contributions so far by different researchers in the respective domains.

10.2.1 Network Congestion

Network Congestion is considered as one of the critical issues and a prominent research area in WSNs. It arises in sensor networks due to the traffic originating via many to one communication between nodes and the base station (BS). This result in different issues like loss of meaningful information, reduced packet delivery ratio, and also give rise to unwanted delay in data communication. In consequence, it profoundly impacts network energy efficiency as well as network lifespan. Several congestion control mechanisms have been presented in literature till date [2–8]. There are basically two kinds of congestion that a network mainly experiences. These types are node level congestion and sink level congestion. Techniques for controlling congestion has been categorized into categories based on four parameters. These parameters are (i) minimizing frequent transmission of data packets; (ii) efficient use of network resources; (iii) adjusting the queue length; and (iv) incorporating priority aware approach in the network. Few authors also suggested two variations in control congestion approach. In the first variation, mobile sensor nodes build local alternative paths heading toward the sink [4–6]. On contrary, in the second variation mobile nodes create their own individual paths to the sink. Simulation outcomes depict that both suggested approaches performed well to alleviate congestion in a significant way [7, 8]. Another main reason for congestion in the network is it occurs when number of packets delivered to the node is more than the number of packets forwarded by that node.

10.2.2 Node Deployment

Appropriate and strategic node deployment in WSN is very essential as it help in tackling with several issues like data routing, data aggregation, and data communication. The importance of this factor is explained from all aspects of networking by different researchers [9–13]. There are basically three different schemes for node deployment. These schemes are uniformly random deployment, square grid-based deployment, and a pattern-based hexagonal node deployment. To make comparison among these three deployment strategies, author adopted few performance metrics namely (i) energy consumption per round, (ii) data communication delay, and (iii) network coverage. Moreover, trade-offs between these metrics are also figured to prove the preferable type of node deployment. Simulation outcomes have proved that a pattern-based deployment is the most promising deployment approach [9–11].

Different node deployment strategies are defined by researchers based on different optimization approaches. Bio-inspired techniques like particle swarm optimization and ant colony optimization are also adopted by several researchers for fair node deployment [12]. Both low cost and high connectivity factors are adopted for an appropriate node deployment scheme. This proposed scheme uses a novel ACO-greedy approach for node distribution in the network [13]. In addition, this scheme also deals with energy hole problem and results in network lifetime enhancement.

A hierarchical-logic based framework can also be employed to deploy sensor nodes in the target region. Disk communication model and radio frequency (RF) propagation model can be utilized to ensure high network connectivity and coverage [13]. Furthermore, an image processing algorithm can be incorporated in such schemes to classify deployment terrain for high network connectivity.

10.2.3 Data Aggregation

It is a mechanism of agglomerating data sensed by SNs in the network. On contrary, data fusion is the process of combining data sensed by the nodes via signal processing. Data fusion works by following few steps. Firstly, it merges signals obtained from the sensor nodes, then it eliminates noise from the merged signals followed by applying some technique to only combine useful and meaningful data. At the end, it generates an accurate signal as the output. Aggregating data before relaying it to the BS has many fold advantages. Since SNs are generally set up in close vicinity to each other, it is very much obvious that multiple nodes may sense same data about a given physical condition. Also, data aggregation reduces number of data transmissions done for passing data to the central authority and hence lessens the overall power consumption. An appropriate data aggregation should focus on three factors namely (i) location of aggregation points; (ii) function derived for aggregating data; and (iii) number of sensors in the target area.

Researchers' interest in designing data aggregating techniques has grown rapidly in past few years [14–21]. Few authors presented an exact algorithms and approximate algorithms to select optimal number of aggregation points in the network. The objective is to enhance network lifetime. Simulation results have proved that approximate algorithms outperform exact algorithms in enhancing network lifespan [14]. Also, one of the authors presented a framework for gathering data in the sensor network by employing aggregation algorithm named Tiny Algorithm (TAG). This data collection strategy proved significantly better than its counterpart. Author so far has claimed that through data aggregation in WSNs, high energy efficiency can be achieved. In [19] author presented previous and current research contributions in this domain and also covers future directions in data aggregation for WSNs. Different existing energy efficient approaches are grid-based data aggregation, temporal correlation-based data aggregation, polygon regression based, and Steiner tree-based data aggregation. These data aggregation approaches help in minimizing network's total energy consumption.

A new secure and energy efficient data aggregation scheme named CSDA [20] was introduced in literature. This scheme makes use of slice assemble approach to achieve this. In addition, this scheme not only minimizes communication overhead and energy utilization meanwhile it also preserves privacy in data communication. A comb needle discovery model for query processing is extended by introducing a cluster-based data combining method into it [21]. This extension results in minimizing the communication cost and energy utilization of the network, thereby enhancing the network life duration.

10.2.4 Network Coverage

Achieving high network coverage is an important issue. The term network coverage explains how efficiently sensor nodes are monitoring the network area [22–25]. High coverage rate promotes high Quality of Service (QoS) in WSNs. WSNs undergo many serious issues. One of such issues arises when the complete energy of a particular node gets drained and it stops functioning. Consequently, coverage rate of that region will gradually decrease resulting in poor network performance. Similar to this issue, there is one more challenge occurring in such scenarios and that challenge is termed as "border effect." Since sensor nodes are battery-operated devices and have limited power availability, they have very less sensing range. An event can be detected by it only if that event lies in its sensing area or in its communication range. The sensing area of a node will be more if it lies near the central region but will be less if it lies on the border of the targeted area. This phenomenon is termed as border effect. So, coverage also relies on the kind of nodes distribution within the network. In consequence, there is an urgent need to introduce some coverage preserving routing protocols.

There exist several network sensing models like (i) Boolean sensing model; (ii) shadow fading model; (iii) Elfes sensing model. It was investigated that the impact of these models on the network coverage is significantly very good in terms of energy conservation through high network coverage. Few researchers have also presented an algorithm to optimize network coverage using swarm intelligence optimizer PSO (Particle Swarm Optimization) and Voronoi diagram [23]. The proposed algorithm guaranteed that the region of interest will be sensed completely by the sensor nodes to detect events. A Simulated Kalman Filter (SKF) has been adopted by a researcher to achieve better coverage than other metaheuristic approaches like GA and PSO [24]. To measure the network coverage and to validate the efficiency of SKF, binary sensing model is incorporated.

10.2.5 Mobility in Wireless Sensor Networks

SNs deployed in WSN can be either mobile or static in nature. These days, Mobile Wireless Sensor networks (MWSNs) are proving to be very effective in different domains like underwater monitoring, target or object tracking, and for e-health monitoring of serious or old-aged patients. Mobility factor is introduced in sensor nodes by author to tackle hotspot problem in WSNs [26–30]. Mobility-based communication results in high connectivity among the sensor nodes. Author categorized two types of data collection depending on sink mobility. These are MDC (Mobile Data Collector) and MBS (Mobile Base Station). Moreover, author also introduces a new approach to evaluate trajectories for mobile data collector [27]. A mobility model to assure controlled disconnections and soft handoffs is introduced by researchers.

Two mobility aware clustering schemes named MCCA "Mobility Aware Centralized Clustering Algorithm" and MHCA "Mobility Aware Hybrid Clustering Algorithm" are proposed in literature. Both these schemes result in stable cluster formation and reduced energy consumption [30]. Also, it improves the network stability of mobile sensor networks. Moreover, author focuses on the adequate choice of CH nodes during cluster formation process.

10.2.6 Base Station Positioning

The bulk of research has been carried out in the designing of optimal base station positing schemes. Base station positioning has significant impact on network coverage area, fault tolerance, energy utilization, and network congestion [31–33]. Base station can be positioned stationary or mobile. Instead of keeping the base station static, the network performance can be enhanced by using dynamic schemes to position BS within the network. In clustered framework with statically positioned BS, CH nodes use either single hop or multi-hop communication to relay their data to the BS [31]. This causes a lot of energy dissipation in these head nodes. On contrary, if the BS is dynamic in nature, then it will keep on moving from one region to the other region within the network, to collect data itself from the head nodes [32]. Author claimed that the dynamic positioning of a BS optimizes both functional and non-functional objectives of network in comparison to the static placement of the BS. A Multi-Objective Metric (MOM) to search out optimal position of the base station has been introduced in literature [33]. This metric focuses on the achievement of few factors like (i) fault tolerance in the network; (ii) nodes distance from the sink; (iii) delay due to network congestion; and (iv) energy consumption.

Also, one of the researchers introduced a metric aware BS placement and relocation procedure for WSNs [34]. Author claimed to solve the BS positioning problem in nonlinear environment by considering path loss exponent value. In this scheme, sink node itself evaluates its position in every round with respect to the positions of the operating nodes. A scheme to choose optimal position for BS based on K-means ++ and local + methods is also highlighted by several other researchers working in this domain. In such schemes, multiple base stations are placed in the target region to cope with energy hole issue. Instead of the nodes communicating their data to a single BS, multiple base stations are deployed so that nodes can relay their sensed data periodically and easily to their nearest BS. This approach results in high energy conservation. In the nutshell, after having a brief review of all these research areas we can say that all these issues are in some respect dependent on each other.

10.2.7 Trusted Communication

Considering deployment of sensors in harsh and unattended regions for monitoring purpose, their vulnerability to severe attacks from the intruders increases more likely. Hence, it is very much essential to introduce some security measures in WSN architecture especially in case of critical applications like military surveillance, health monitoring, etc. Moreover, in clustered network, cluster head node keeps all confidential and important data collected from its cluster member. Hence, the selection of trustworthy nodes as cluster head is also very much necessary.

Few of the existing algorithms are discussed in this section [34–37]. One of the researchers proposed a distributed trust-based framework to select trustworthy nodes as leader nodes. This trust model deals with bad mouthing attack because nodes are supposed to share their computed trust information about other nodes with their own head nodes only. Hence, no communication between cluster members took place. A trust based secure clustering scheme named TREE-CR was devised [35]. In this scheme, author has discussed two phases for designing a trusted framework for routing in WSNs. These phases are (i) cluster head node selection and (ii) cluster establishment phase. Furthermore, to predict accurate lifetime of network, a novel and realistic power utilization model was introduced. A new scheme named LWTC-BMA was introduced to extend TREE-CR protocol. LWTC-BMA [36] aims on enhancing the network lifetime by expelling malicious nodes from being chosen as a leader node. Unlike TREE-CR, this approach evaluates node's trust by adopting a nature inspired algorithm named Honey Bee Mating Optimization (HBMO). HBMO selects two types of head nodes (i) premier phase CHs and (ii) surrogate CHs among the nodes within the cluster.

WOATCA [37] is a novel aware energy efficient protocol based on Whale Optimization Algorithm (WOA). This scheme makes use of five parameters including trust parameters for the selection of CH nodes. These parameters are (i) node's remaining power level; (ii) number of packets relayed by a node to other nodes; (iii) average cluster distance; (iv) delay in transmission activity; and (v) node density into an account. This approach tries to expel malicious nodes from being elected as CH. eeTMFO/GA is a trust-based clustering scheme designed using Moth Flame Optimization (MFO) and Genetic Algorithm (GA). This scheme focuses on the selection of trustworthy nodes as head nodes and avoids the selection of malicious nodes for this role. This selection is done using few important parameters. These parameters are (i) power level; (ii) count of successfully delivered packets; (iii) distance; (iv) data transmission delay activity; and (v) node count. Simulation results validate efficient working of this proposed scheme. Author in [38] discussed analytical modeling based on various trust-based protocols.

10.2.8 Time Synchronization

Time synchronization among sensors is highly desirable to ensure accurate network operations. Numerous protocols have been proposed till date for time synchronization in WSNs [39–43]. Network activities that demand proper time synchronization among the nodes are (i) data aggregation; (ii) TDMA scheduling; and (iii) node's status updation like setting node from active mode to sleep state or sleep mode to receive mode. A lightweight time synchronization scheme named TSync was introduced. This scheme aims on providing time synchronization on global as well as individual node basis. The simulation outcomes for this approach shows accurate time stamping of WSN operations and aids in fine tuning of wake sleep duty cycles. A comprehensive study to evaluate existing clock synchronization schemes in WSN based on multiple factors like cost, accuracy, precision, and complexity are provided. A novel clock synchronization scheme named RTMS focusing primarily on bounded noises exists in the literature. The objective of this scheme is to improve accuracy in clock synchronization through minimization in communication delay and packet loss rate.

10.2.9 Avoiding Energy Holes in WSNs

Energy hole problem is also one of the critical challenges experienced by sensor networks. This issue arises due to heavy traffic load carried by the nodes near the BS. During the data communication from the nodes positioned far away from the BS, multi-hop communication is adopted for data forwarding. Source nodes select other nodes as intermediate nodes termed as gateways, to route their data to the BS. Due to this, the energy dissipated by the nodes located nearer to the BS are comparatively more in comparison to the other nodes. Hence, these nodes die earlier resulting in energy holes around the BS. To cope with this issue, load balancing among the nodes is very much essential [44].

A Grey Wolf Optimization (GWO) approach based accurate node clustering is proposed to distribute the load evenly among the gateways [45]. Two novel fitness functions are derived one for each clustering and routing purpose. The fitness function is drafted in such a manner that the number of hops and the traversal distance among the nodes are minimized to achieve even load distribution. Author tries to deal with energy hole problem through the adoption of non-uniform node distribution strategy. This proposed approach results in balanced energy dissipation within the network.

10.2.10   TDMA Scheduling

TDMA scheduling refers to the allocation of slots to each node in the network for the conflict free data transmission. One of the authors introduced two centralized algorithms based on heuristics to prove that TDMA scheduling is NP complete problem [46]. First algorithm adopts node-based scheduling and second uses level-based scheduling for many to one communication in WSNs. TDMA sleep scheduling is another approach to minimize node's energy consumption. By following this approach, nodes undergo multiple state transitions during the complete network lifetime. A sensor node can switch between multiple states. These states are idle state, receiving state, transmitting state, and sleep state. TDMA sleep scheduling sends the node in sleep state when it has no packets to forward. This helps in avoiding the energy consumed by node otherwise spent during idle listening. A novel conflict avoiding TDMA sleep scheduling algorithm was also presented that minimize the frequent changes in state transition. This approach results in energy conservation of SNs [47].

10.2.11   Fault Tolerance

Deployment of sensor nodes in harsh environments may lead to severe faults at any time in the network. These faults can be due to sink failure, node failure, or failure in network operations. Faults occurring in the network are a clear depiction of network reliability. Classification of different types of attacks and survey about fault tolerance routing schemes in sensor networks exists in literature [48–50]. A mechanism to recover failed sensor nodes in a cluster is introduced in this scheme. This fault recovery scheme focuses on avoiding the deployment of redundant gateways in the network region as well as it avoids the overhead caused due to re-clustering process.

10.2.12   Load Balancing among Nodes

Clustered WSNs consists of two fundamental phases (i) selecting CH nodes and (ii) assigning sensor nodes to these CHs. Grouping the network into clusters helps in improving the network scalability. This grouping includes the selection of few head nodes to communicate with the central authority BS instead of every single node communicating with the BS. But this approach requires the distribution of head nodes within a network in a well-balanced manner. If these head nodes are not well distributed then it would result in uneven load distribution among them. The head nodes located in highly dense area will be having more nodes associated to them and hence their energy will get drained quickly. Consequently, this would result in degradation in the network performance. Several researchers have presented load balanced clustering approaches for uniform load distribution among the sensor nodes [51–57].

A load balanced clustering algorithm for sensor networks was proposed by a researcher [58, 59]. This algorithm assumes the deployment of nodes with less energy restrictions as gateways along with high energy restricted nodes as normal sensors. These gateway nodes act as CHs. The objective of the proposed scheme is to balance workload among these gateways. Simulation results depict how this scheme results in accurate balancing of load thereby improving the network lifetime. A load-balancing clustering scheme is introduced for randomly distributed nodes in the network. This algorithm uses node's energy and distribution density to cluster the network. Simulation outcomes have shown achievement in network lifetime longevity. An approach named Energy Efficient Load-Balanced Clustering (EELBC) [51] is introduced to balance load and to address energy efficiency issue in sensor networks.

The proposed algorithm EELBC involves minimum heap-based clustering. A minimum heap is built of cluster head nodes along with their assigned cluster members. This algorithm has a time complexity of O (n log m) for n sensor nodes and m CHs. The simulation results proved that EELBC performs very well with respect to energy efficiency, load balanced distribution, and number of dead nodes per round. Author utilizes Away Cluster Head (ACH) scheme to cluster sensor networks. This scheme minimizes uneven load distribution and back transmission among nodes. The head nodes are selected based on two factors i.e., remaining energy and distance from other CH nodes. An additional phase named free association is adopted to associate the nodes into cluster that helps in avoiding the back transmission. This proposed algorithm results in improved network lifetime and energy conservation.

One of the authors focuses on the load balancing of cluster head nodes by using an evolutionary approach named Genetic Algorithm (GA) [53]. The proposed scheme performs well in case of even as well as uneven load distribution among the nodes. Simulation outcomes show that this scheme performs significantly well in terms of metrics like algorithm run time, energy dissipation, number of alive nodes and alive CHs, load balancing, and convergence rate. Also, a Load Balanced Clustering Problem (LBCP) was introduced in literature. In this attempt, network is grouped into clusters to improve network scalability. The objective of this work is to balance load among head nodes called gateways. This load balancing approach improves communication among the nodes as well as enhances network stability. In addition, author claimed that special case of LBCP can be can be solved in polynomial time and generally it is NP-hard problem. Figure 10.1 below represents the use case diagram that highlights the dependency of these research domains with more clarity.

WSNs is an important and necessary technology for the future of tremendously adopted concept i.e., IOT "Internet of Things." Sensors are the devices that are used for the purpose of monitoring and controlling in several domains. WSNs transmit sensed data and control signals over large distances without the need of expensive underlying infrastructure. However, the quality of data sensed by these sensors is affected by

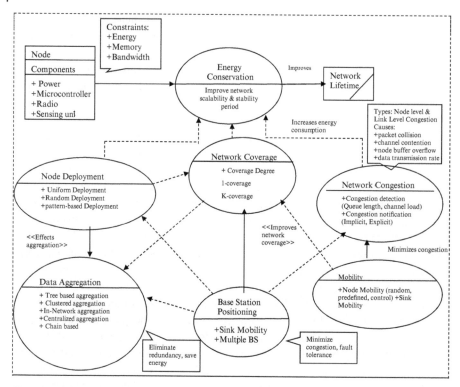

Figure 10.1 Use case diagram showing dependency among different challenges in WSNs.

challenges like sensing device failures, environment changes, or due to malicious attacks. Due to these limitations of sensor networks, there is a need to contribute in the above discussed research areas to efficiently make use of these sensor devices.

10.3 Application Areas of Advanced WSNs

The research dimensions and application areas of WSNs are advancing day by day. Few of the application areas are:

a) **Structural Health Monitoring (SHM):** SHM includes linking of different kinds of sensor devices to collect the desired information about the health of the structures. This helps in reducing the further risk to them by detecting the localized damage and predicting the network life span of the structure. Bridge monitoring is one of the most prominent applications of SHM.

b) **Internet of Things (IoT):** The integration of WSN with IoT has lead to development of smart environmentthe development of smart cities, smart homes, smart offices, etc. In such environment sensors make use of Internet services in a dynamic fashion to work collaboratively.

c) **Surveillance and Monitoring for Security and Threat Detection:** Advancement in WSNs has resulted in better surveillance and monitoring of crowded places to ensure more security of assets. Making use of sensors at home or work premises is an ideal solution for wide range surveillance.

d) **Disaster Recovery and Relief Operations:** These days sensors are operated in areas affected due to catastrophic situations for disaster recovery and relief operations. For landslide detection also sensors are deployed in harsh and unattended regions.

e) **Agriculture Monitoring:** In agriculture domain, physical sensors are used these days to monitor water quality, soil moisture content. The management of crop cultivation is done using sensors that helps in predicting the exact condition in which plants should be grown and to find out the optimum conditions for every crop based on past experiences.

f) **Medical Applications like Patient Monitoring:** WSNs can be used for long-term real-time monitoring of chronologically ill patients.

10.4 Conclusion

With the advancement in wireless technology, sensor nodes can leverage the strength of working collaboratively to achieve high quality wireless communication. Nowadays, sensors and WSNs have become an integral part of our lives. However, due to the resource restricted nature of these sensing devices, we are facing several challenges in deploying them. This chapter identified several research areas like node deployment, data aggregation, clustering, and base station positioning that needs further study. This chapter will inspire and guide researchers to further advance WSNs and their applications in practice.

Bibliography

1 Akyildiz, I.F., Su, W., Sankarasubramaniam, Y., and Cayirci, E. (2002). Wireless sensor networks: a survey. *Computer Networks* 38 (4): 393–422.

2 Dashkova, E. and Gurtov, A. (2012). Survey on congestion control mechanisms for wireless sensor networks. In: *Internet of Things, Smart Spaces, and Next Generation Networking*, 75–85. Berlin, Heidelberg: Springer.

3 Hashemzehi, R., Nourm, R., and Koroupi, F. (2013). Congestion in wireless sensor networks and mechanisms for controling congestion. http://citeseerx.ist.psu.edu/viewdoc/summary? doi: 10.1.1.375.4567.

4 Ghaffari, A. (2015). Congestion control mechanisms in wireless sensor networks: a survey. *Journal of Network and Computer Applications* 52: 101–115.

5 Shah, S. A., Nazir, B., and Khan, I. A. (2017). Congestion control algorithms in wireless sensor networks: trends and opportunities. *Journal of King Saud University-Computer and Information Sciences* 29 (3): 236–245.

6 Tabatabaei, S. and Omrani, M. R. (2018). Proposing a method for controlling congestion in wireless sensor networks using comparative fuzzy logic. *Wireless Personal Communications* 100 (4): 1459–1476.

7 Nicolaou, A., Temene, N., Sergiou, C., Georgiou, C., and Vassiliou, V. (2019). Utilizing mobile nodes for congestion control in wireless sensor networks. arXiv preprint arXiv:1903.08989.

8 Kazmi, H. S. Z., Javaid, N., Imran, M., and Outay, F. (2019). Congestion control in wireless sensor networks based on support vector machine, grey wolf optimization and differential evolution. *2019 Wireless Days (WD)*, April, 1–8. IEEE.

9 Poe, W. Y. and Schmitt, J. B. (2009). Node deployment in large wireless sensor networks: coverage, energy consumption, and worst-case delay. *Asian Internet Engineering Conference*, November, 77–84. ACM.

10 Li, Z. and Lei, L. (2009). Sensor node deployment in wireless sensor networks based on improved particle swarm optimization. *2009 International Conference on Applied Superconductivity and Electromagnetic Devices*, September, 215–217. IEEE.

11 Zhang, H. and Liu, C. (2012). A review on node deployment of wireless sensor network. *International Journal of Computer Science Issues (IJCSI)* 9 (6): 378.

12 Liu, X. and He, D. (2014). Ant colony optimization with greedy migration mechanism for node deployment in wireless sensor networks. *Journal of Network and Computer Applications* 39: 310–318.

13 Olasupo, T. O. and Otero, C. E. (2018). A framework for optimizing the deployment of wireless sensor networks. *IEEE Transactions on Network and Service Management* 15 (3): 1105–1118.

14 Al-Karaki, J. N., Ul-Mustafa, R., and Kamal, A. E. (2004), April. Data aggregation in wireless sensor networks-exact and approximate algorithms. *2004 Workshop on High Performance Switching and Routing*, April 2004, HPSR, 241–245. IEEE.

15 Nandini, S. P. and Patil, P. R. (2010). Data aggregation in wireless sensor network. *IEEE International Conference on Computational Intelligence and Computing Research*, 1–6.

16 Dagar, M. and Mahajan, S. (2013). Data aggregation in wireless sensor network: a survey. *International Journal of Information and Computation Technology* 3 (3): 167–174.

17 Randhawa, S. and Jain, S. (2017). Data aggregation in wireless sensor networks: previous research, current status and future directions. *Wireless Personal Communications* 97 (3): 3355–3425.

18 Dhand, G. and Tyagi, S. S. (2016). Data aggregation techniques in WSN: survey. *Procedia Computer Science* 92: 378–384.

19 Gherbi, C., Aliouat, Z., and Benmohammed, M. (2015). Distributed energy efficient adaptive clustering protocol with data gathering for large scale wireless sensor networks. *2015 12th International Symposium on Programming and Systems (ISPS)*, April, 1–7. IEEE.

20 Fang, W., Wen, X., Xu, J., and Zhu, J. (2019). CSDA: a novel cluster-based secure data aggregation scheme for WSNs. *Cluster Computing* 22 (3): 5233–5244.

21 Shanmukhi, M. and Ramanaiah, O. B. V. (2015). Cluster-based comb-needle model for energy efficient data aggregation in wireless sensor networks. *2015 Applications and Innovations in Mobile Computing (AIMoC)*, February, 42–47. IEEE.

22 Hossain, A., Biswas, P. K., and Chakrabarti, S. (2008). Sensing models and its impact on network coverage in wireless sensor network. *2008 IEEE Region 10 and the Third international Conference on Industrial and Information Systems*, December, 1–5. IEEE.

23 Ab Aziz, N. A. B., Mohemmed, A. W., and Alias, M. Y. (2009). A wireless sensor network coverage optimization algorithm based on particle swarm optimization and

Voronoi diagram. *2009 International Conference on Networking, Sensing and Control*, March, 602–607. IEEE.

24 Aziz, N. A. A., Ibrahim, Z., Aziz, N. H. A., and Aziz, K. A. (2019). Simulated Kalman filter optimization algorithm for maximization of wireless sensor networks coverage. *2019 International Conference on Computer and Information Sciences (ICCIS)*, April, 1–6. IEEE.

25 Kong, H. and Yu, B. (2019). An improved method of WSN coverage based on enhanced PSO algorithm. *2019 IEEE 8th Joint International Information Technology and Artificial Intelligence Conference (ITAIC)*, May, 1294–1297. IEEE.

26 Basagni, S., Carosi, A., Petrioli, C., and Boukerche, A. (2008). *Mobility in Wireless Sensor Networks*. Wiley Series on Parallel and Distributed Computing, 267–305. Hoboken, NJ: John Wiley & Sons, Inc.

27 Ekici, E., Gu, Y., and Bozdag, D. (2006). Mobility-based communication in wireless sensor networks. *IEEE Communications Magazine* 44 (7): 56–62.

28 Silva, R., Zinonos, Z., Silva, J. S., and Vassiliou, V. (2011). Mobility in WSNs for critical applications. *2011 IEEE Symposium on Computers and Communications (ISCC)*, June, 451–456. IEEE.

29 Zafar, S., Bashir, A., and Chaudhry, S.A. (2019). Mobility-aware hierarchical clustering in mobile wireless sensor networks. *IEEE Access* 7: 20394–20403.

30 Tolba, F. D., Ajib, W., and Obaid, A. (2013). Distributed clustering algorithm for mobile wireless sensors networks. SENSORS, November, 1–4. IEEE.

31 Akkaya, K., Younis, M., and Youssef, W. (2007). Positioning of base stations in wireless sensor networks. *IEEE Communications Magazine* 45 (4): 96–102.

32 Kim, S., Ko, J. G., Yoon, J., and Lee, H. (2007). Multiple-objective metric for placing multiple base stations in wireless sensor networks. *2007 2nd International Symposium on Wireless Pervasive Computing*, February. IEEE.

33 Zadeh, P. H., Schlegel, C., and MacGregor, M. H. (2012). Distributed optimal dynamic base station positioning in wireless sensor networks. *Computer Networks* 56 (1): 34–49.

34 Crosby, G. V., Pissinou, N., and Gadze, J. (2006). A framework for trust-based cluster head election in wireless sensor networks, 13–22.

35 Sahoo, R. R., Singh, M., Sardar, A. R., Mohapatra, S., and Sarkar, S. K. (2013). TREE-CR: trust based secure and energy efficient clustering in WSN. *Emerging Trends in Computing, Communication and Nanotechnology (ICE-CCN), 2013 International Conference on March*, 532–538. IEEE.

36 Sahoo, R. R., Singh, M., Sahoo, B. M., Majumder, K., Ray, S., and Sarkar, S. K. (2013). A light weight trust based secure and energy efficient clustering in wireless sensor network: honey bee mating intelligence approach. *Procedia Technology* 10: 515–523.

37 Sharma, R., Vashisht, V., and Singh, U. (2020). WOATCA: whale optimization algorithm based trusted scheme for cluster head selection in wireless sensor networks. *IET Communications* 14 (8): 1199–1208.

38 Sichitiu, M. L. and Veerarittiphan, C. (2003). Simple, accurate time synchronization for wireless sensor networks. *2003 IEEE Wireless Communications and Networking, WCNC 2003*, Vol. 2, March, 1266–1273. IEEE.

39 Dai, H. and Han, R. (2004). TSync: a lightweight bidirectional time synchronization service for wireless sensor networks. *ACM SIGMOBILE Mobile Computing and Communications Review* 8 (1): 125–139.

40 Sundararaman, B., Buy, U., and Kshemkalyani, A. D. (2004). Clock synchronization for wireless sensor networks: a survey. *Ad Hoc Networks* 3 (3): 281–323.

41 Rhee, I. K., Lee, J., Kim, J., Serpedin, E., and Wu, Y.C. (2009). Clock synchronization in wireless sensor networks: an overview. *Sensors* 9 (1): 56–85.

42 Zhang, X., Chen, H., Lin, K., Wang, Z., Yu, J., and Shi, L. (2019). RMTS: a robust clock synchronization scheme for wireless sensor networks. *Journal of Network and Computer Applications* 135: 1–10.

43 Lipare, A., Edla, D. R., and Kuppili, V. (2019). Energy efficient load balancing approach for avoiding energy hole problem in WSN using Grey Wolf Optimizer with novel fitness function. *Applied Soft Computing* 84: 105706.

44 Wu, X., Chen, G., and Das, S. K. (2008). Avoiding energy holes in wireless sensor networks with nonuniform node distribution. *IEEE Transactions on Parallel and Distributed Systems* 19 (5): 710–720.

45 Ergen, S. C. and Varaiya, P. (2010). TDMA scheduling algorithms for wireless sensor networks. *Wireless Networks* 16 (4): 985–997.

46 Ma, J., Lou, W., Wu, Y., Li, X. Y., and Chen, G. (2009). Energy efficient TDMA sleep scheduling in wireless sensor networks. *IEEE INFOCOM*, April, 630–638. IEEE.

47 De Souza, L. M. S., Vogt, H., and Beigl, M. (2007). *A Survey on Fault Tolerance in Wireless Sensor Networks*. Interner Bericht. Fakultät für Informatik, Universität Karlsruhe.

48 Alwan, H. and Agarwal, A. (2009). A survey on fault tolerant routing techniques in wireless sensor networks. *2009 Third International Conference on Sensor Technologies and Applications*, June 2009, 366–371. IEEE.

49 Gupta, G. and Younis, M. (2003). Fault-tolerant clustering of wireless sensor networks. *2003 IEEE Wireless Communications and Networking, 2003. WCNC 2003*, Vol. 3, March 2003, 1579–1584. IEEE.

50 Gupta, G. and Younis, M. (2003). Load-balanced clustering of wireless sensor networks. *IEEE International Conference on Communications, 2003. ICC'03*, Vol. 3, May, 1848–1852. IEEE.

51 Liao, Y., Qi, H., and Li, W. (2012). Load-balanced clustering algorithm with distributed self-organization for wireless sensor networks. *IEEE Sensors Journal* 13 (5): 1498–1506.

52 Kuila, P. and Jana, P. K. (2012). Energy efficient load-balanced clustering algorithm for wireless sensor networks. *Procedia Technology* 6: 771–777.

53 Sharma, A. and Kansal, P. (2015). Energy efficient load-balanced clustering algorithm for wireless sensor network. *2015 Annual IEEE India Conference (INDICON)*, December, 1–6. IEEE.

54 Kuila, P., Gupta, S. K., and Jana, P. K. (2013). A novel evolutionary approach for load balanced clustering problem for wireless sensor networks. *Swarm and Evolutionary Computation* 12: 48–56.

55 Low, C. P., Fang, C., Ng, J. M., and Ang, Y. H. (2008). Efficient load-balanced clustering algorithms for wireless sensor networks. *Computer Communications* 31 (4): 750–759.

56 Sharma, R., Vashisht, V., and Singh, U. (May 2019). EEFCM-DE: energy efficient clustering based on fuzzy C means and differential evolution algorithm in wireless sensor networks. *IET Communications* 13 (8): 996–1007.

57 Sharma, R., Vashisht, V., and Singh, U. (2019). eeFFA/DE- A fuzzy based clustering algorithm using hybrid technique for wireless sensor networks. *International Journal of Artificial Intelligence Paradigms.* doi: 10.1504/IJAIP.2019.10025734.

58 Sharma, R., Vashisht, V., and Singh, U. (2019). Fuzzy modelling based energy aware clustering in wireless sensor networks using modified invasive weed optimization. *Journal of King Saud University – Computer and Information Sciences.* doi: 10.1016/j. jksuci.2019.11.014.

59 Sharma, R., Vashisht, V., and Singh, U. (2020). Analytical modeling of trust based protocols for cluster head selection in wireless sensor networks. *2020 8th International Conference on Reliability, Infocom Technologies and Optimization (Trends and Future Directions) (ICRITO)*, 643–647. doi: 10.1109/ICRITO48877.2020. 9197913.

11

Synergizing Blockchain, IoT, and AI with VANET for Intelligent Transport Solutions

S.S. Zalte[1], V.R. Ghorpade[2] and Rajanish K. Kamat[3]

[1] Assistant Professor in Computer Science Department at Shivaji University, Kolhapur, India
[2] Ph. D. in Computer Science and Engineering from Shri Guru Govindsinghji Institute of Engineering and Technology (An Autonomous Institute of Govt. of Maharashtra), Nanded
[3] Dean, Faculty of Science & Technology in addition to Professor in Electronics and Head of the Department of Computer Science at Shivaji University, Kolhapur a NAAC A++ accredited HEI

11.1 Introduction

The rapid adoption and development of mobile devices have enabled ubiquitous computing, which has opened up new research areas such as smart cities, intelligent transportation, mobile health, and remote monitoring. Recently, the Kaspersky Lab, a global research and analysis team, explores safety issues in smart cities and transport infrastructure and recommendations on how to address them. A smart city scenario is an innovation, challenge, and opportunity where information and communication technologies (ICT) are used to serve people and improve the economy, environment, governance, life, and mobility. Smart cities are often referred to as digital or connected cities as they involve the smart use of technology to add value, deliver more efficient services, and alleviate the growing urbanization and population problems. Citizens interact with smart city services via their smartphones and computers connected to a heterogeneous network of systems. This approach is made possible by providing intelligence through mobile sensors and wireless capabilities in infrastructures such as green buildings, portable devices, vehicles and intelligent transport systems to facilitate data access, essential for realizing smart cities.

Simply put, smart cities rely on connected devices to streamline and improve urban services based on extensive real-time data. These devices combine hardware, software and geo-analytics with improving community services and improving areas of life. The objects and connected IoT used in smart cities are characterized by ubiquity, miniaturization, autonomy, unpredictable behavior and difficult identification. With billions of devices getting connected worldwide, the governing bodies of smart cities are increasingly concerned about the security aspects. Solutions to address the security and privacy concerns of smart cities based on different smart objects fall not only in technology but also in other fields such as sociology, law and policy management.

Thus as reviewed above, faced with swift population growth, in the wake of smart cities, effective, sustainable, and secure solutions to transport, climate, energy, and governance are the need of the hour. Smart cities also necessitate intelligent transportation, and many innovative technological platforms are emerging to facilitate the same. One such proven platform are Vehicular Ad-hoc Network (VANET). VANET was created to facilitate communication between the vehicles themselves

Emerging Computing Paradigms: Principles, Advances and Applications, First Edition.
Edited by Umang Singh, San Murugesan and Ashish Seth.
© 2022 John Wiley & Sons Ltd. Published 2022 by John Wiley & Sons Ltd.

and the vehicle infrastructure. It turns participating cars into wireless router nodes, allowing cars to connect within 100–300 meters, creating a long-range network.

Nevertheless, another technology smoothing the transactions in general and in the specific context of smart cities is blockchain. Without any doubt, the blockchain enables individuals and organizations to conduct immediate and secure transactions over a distributed network. A new combination of sophisticated technological concepts, including peer-to-peer networks, distributed consensus algorithms, validity rules, ledger technology, cryptography and blockchains, can be applied to any area where trust-based relationships are disrupted. From a general perspective, anyone who has followed banking, investment, or cryptocurrencies over the past decade is probably familiar with the term "blockchain"—the recording technology behind the Bitcoin network. It is regarded as the most important technological innovation since the Internet or the solution to the most forward-looking problems, VANET is just one.

The present chapter focuses on reviewing the analytical marriage of blockchain and Vehicular Ad-hoc Network (VANET) from security. As with any distributed database, each party to the blockchain has access to the entire database and its entire history. Moreover, each party may review the records of its transactions with a partner or intermediary. Moreover, communication takes place between peers without a central node. Given this, security plays a very crucial role in applications such as VANET. First, the chapter put forth the said security concerns through a systematic review of the blockchain technology followed by the VANET. After that, the security concerns and research work thereof is presented. In the end, the future directions are indicated.

11.2 Blockchain: Growing beyond the Traditional Crypto currency Paradigm

Blockchain is the buzzword that acquires more attention. In any case, a large portion of us is not mindful of what blockchain is or how to depict it to other people. Generally, people imagine that blockchain is Bitcoin and the other way around. In any case, it is not the situation. Bitcoin is a digital currency blockchain developed by Satoshi Nakamoto. As the name proposes, blockchain is a chain of blocks that contains data. Each block comprises various transactions, and every transaction is recorded as Hash. Hash is a fixed-length value allocated to each block during its creation, and any further alteration in the block will prompt an adjustment in its hash. Thus, *blockchain* is a technology that contains a record list of transactions made digital currency like Bitcoin, etherum or another cryptocurrency are maintained over distributed several computers linked in a peer-to-peer network.

The blockchain has gained so much attention because of its popular characteristics as given below:

- **Decentralization:** No centralized authority multiple entities involved
- **Encrypted:** Scrambled data stored inside the blockchain by applying the cryptographical algorithm.
- **Tamperproof:** No one can alter or modify the data inside the blockchain.
- **Transparency:** Every node in the blockchain is aware of transactions made.

The very reason behind the popularity of blockchain technology is its security: blocks added to the are invariable and time-stamped. Further, the Records cannot be manipulated, and the technology does not allow changes to the information stored on the chain. Blockchain is at the core of Bitcoin and other virtual currencies. It has attracted many applications due to its striking feature of an open, distributed register that records transactions between the involved parties verifiably and permanently. Moreover, the inherent peer-to-peer network facilitates tokenization, digital identity, and the transfer of value through decentralized applications. The blockchain originated with Bitcoin, the most widely accepted and widely used cryptocurrency. However, soon the potential of blockchain applications extends far beyond digital currencies such as Bitcoin and other cryptocurrencies, and it made its mark in the Fin-tech industry. Furthermore, emphasizing the integration of decentralized blockchains, the applications are now expanded beyond traditional sectors such as finance, commodity transactions, government supply chain management, smart energy, health, market surveillance, and education.

Many research groups are working on blockchain technology from a different perspective. Zwitter and Hazenberg present the policy implications related to blockchain technology [1]. Xiong et al. [2] review the agricultural applications. Düdder et al. [3] revealed a holistic perspective of applications of blockchain technology out of which those significant from the open science projects are depicted by [4]. Applications of blockchain in the energy sector are picked up as described by [5] while those in food and agriculture from customer perspectives are covered [6]. The other emerging potential application areas are also reported in the literature. A noteworthy research paper in this regards is by [7].

The use of blockchain technology in smart transport applications, especially the platforms like VANET, the topic of deliberation of this chapter, is increasingly supported by the research community and is gaining momentum from an industry perspective. From this viewpoint, it would be worthwhile to look at the types of blockchain and the relevant details.

11.3 Blockchain: Types and Architecture Minutiae

11.3.1 Types of Blockchain

In order to appreciate the blockchain driven VANET system, the types and architecture of the technology need to be put in place. Andreev et al. [8] and Sabry et al. [9] have taken a comprehensive review of the types of blockchain. There are four different major types of blockchain types, as shown in the Figure 11.1. They include the following [8]:

- public blockchain
- private blockchain
- consortium blockchain
- hybrid blockchain

- **Public Blockchain**
  Without any authorization, anyone can quickly join a public chain and do transactions. It is a non-prohibitive adaptation where each node has a duplicate of the

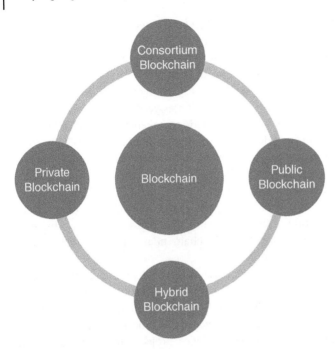

Figure 11.1 Types of blockchain.

ledger. This additionally implies anybody can get to public blockchain on the off chance that they have an Internet connection. Bitcoin is the first type of public blockchain that was delivered to general society. In a decentralized way, anyone can perform transactions through the Internet. Proof-of-Work (PoW) and Proof-of-Stake are significant consensus parameters mechanism used for the validation of transactions. Examples of public blockchain are Bitcoin, Ethereum, Litecoin, and NEO.

- **Private Blockchain**
  A private blockchain operates in a closed environment. Under the authorized entity, only selected nodes can join the private blockchain in a restrictive manner. Hence, it is also called a permission blockchain. In a private organization, this type of blockchain is used for internal use cases. By considering accessibility, authorization, etc., private blockchain providing transparency, trust, and security to the restricted participants.

  Another significant difference is that it is centralized as only one authority looks over the network. So, it does not have a decentralized theoretical nature. Examples of private blockchain are multichain, hyperledger fabric, hyperledger sawtooth, and corda.

- **Consortium Blockchain**
  A consortium blockchain is an inventive way to unravel the necessities of associations where there is a requirement for both public and private blockchain features. In a consortium blockchain, a few parts of the organizations are unveiled, while others stay private. The preset nodes constrain the consensus methodology in a consortium blockchain. A consortium blockchain is overseen by more than one organization. In this way, there is single nobody power brought together result here.

The consortium has a validator node that can help complete two capacities, approve transactions, and start or get transactions to guarantee legitimate usefulness. In correlation, the part node can get or start transactions. So, it offers all the features of a private blockchain, including straightforwardness, protection, and proficiency, without one party having a combining power.

Examples of consortium blockchain are Marco Polo, Energy Web Foundation, and IBM Food Trust.

- **Hybrid Blockchain**
  Hybrid blockchain is an amalgam of a private and public blockchain. It combines advantages of a private and public blockchain. Example of hybrid blockchain are Dragonchain and XinFin's hybrid blockchain.

11.3.2 Cryptography Integration with Blockchain

Security concerns of the blockchain applications cannot be even imagined without the inherent security/cryptography implications. The distributed ledger technology of blockchain gets strengthened in terms of security with the cryptography, the mechanism of scrambling the original plain text into encrypted text (cypher text) and then back into by decrypting cipher text into original plain text. Cryptography helps to secure data from unauthorized access or attackers, more so in the light of domains such as digital identity, platform-as-a-service, and tokenized national currency. In all such applications, confidentiality is the primary concern.

However, given the security requirements, the computationally intensive nature of the cryptographic mechanism sometimes makes the overall performance a significant concern. Cryptography is inherently a building block in providing security primitives like integrity, authentication, and non-repudiation etc. defined on the following lines:

- Confidentiality—Sensitive information cannot reveal in front of an intruder.
- Authentication—It is the process of verifying and confirming users.
- Data Integrity—The cryptographic hash functions play a vital role in assuring the users about data integrity.
- Non-repudiation—It is the process of verifying and confirming a user's identity.

The cryptography implications have evolved the blockchain architecture, the basic notion of which is covered briefly in the following section to appreciate its integration with the VANET.

11.3.3 Overview of Blockchain Architecture

Blockchain is based on peer-to-peer network, where every participant interacts with each other without involving central authority. As it provides transparency, every node is aware of other participants and is copied of data or transactions. Furthermore, each node has a pair of public keys and private keys to preserving nodes' privacy and unique identification. This enriches the blockchain in terms of unique properties like decentralization, transparency, immutability, and security.

The blockchain collects a list of records stored in a specific form of "blocks" linked using cryptographical techniques from the architecture viewpoint. Each block hold hash of its preceding block in encrypted form, timestamp, which shows when the block is created and data related to the transaction. The hash that is the fixed digest of each block is created using a popular hashing algorithm (SHA256). This fixed digest value used in the digital signature for each block works as a fingerprint for the unique identification of the block. It also preserves integrity. If the single bit gets altered in the data, the hash value also changed. Once a record or transaction is stored in the blockchain, no one can be altered or tampered with without leading to change in all the subsequent blocks. This property makes blockchain tamper-proof and immutable. The initial block in the blockchain is called the genesis of the block, which is shown in Figure 11.2. Header field in the blockchain contains metadata, and the body contains transaction data. The relevant details are as follows:

Header part details:

- Prev Hash: It stores fixed digest, i.e., hash value of the preceding block.
- Timestamp: It is used to validate the authenticity, and it also reveals the time instance when the block was used.
- Version: It defines the version of blockchain that is used. There are different versions like version 1.0 (cryptocurrency), 2.0 (smart contract), 3.0 (DAPPS), 4.0 (for industry).
- Random number (Nonce): "number only used once" used match difficulty level restrictions.
- Merkle root: It is a binary tree that holds the hash value of all block transactions used to detect whether data is modified.
- Difficulty Target: It is the level of complexity. With more complexity, miners need to use more powerful and computationally expensive resources called the hash rate of participants.

The network security part of the blockchain is robust because the cryptographic and decentralized methods are embedded. One of the actual depictions of blockchain innovation is that it permits records to be shared over the entire organization and is persistently refreshed appropriately. Thus, the record that every node receives in the appropriated network has an indistinguishable duplicate. At whatever point there is another transaction, the record of this exchange is communicated network-wide and afterwards, different nodes check it in the network. When this transaction is on the whole confirmed, another block is added to the blockchain, and the new network engenders through the network, so every node has a refreshed form.

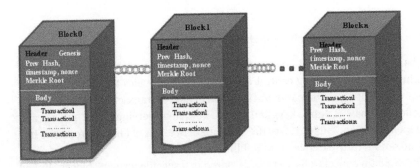

Figure 11.2 Basic structure of blockchain.

Notwithstanding, the way the organization is appropriated agrees to the nodes testing; subsequently, a cryptographic calculation is utilized inside blockchain that depends on the public-private key cryptography. To add or allude to a new block in the blockchain, particular nodes called miners validate the transaction in the newly created block. A block took 10 minutes for mining. Miners have to compete with others to solve a complex puzzle based on a cryptographical hash algorithm. Varying the value of nonce and hash value of nodes, a block is mined until the mathematical puzzle's requirements meet. For solving this mathematical puzzle, miners should have many powerful resources that may be proportional to the hash rate of the node. This process is called proof-of-work. Information regarding newly created block stored in all participants ledger, which various participants manage, is known as Distributed Ledger Technology (DLT).

Proof-of-work and DLT are the key to successful blockchain applications in the VANET. The basic details of VANET is the central focus of the following subpoint. This is then followed by revelation of embedding of blockchain in VANET.

11.4 VANET Based Intelligent Transportation System

Smart transportation systems are at the heart of the ever-growing urban cities. The progress in the sensors, systems, IoT going hand in hand with the operating systems, application software, and Artificial Intelligence (AI) has enabled a holistic transportation system built-in intelligence universally known as smart or intelligent transportation system. The inherent problems of efficiency, safety, and security are addressed through this intelligent transportation system by a mechanism comprising of monitoring, evaluation, and management through the advanced hardware-software protocols. [Smart and secured transportation system using IoT, 10], [11], [guest editorial: intelligent transportation systems in smart cities for sustainable environments (SCfSE), [12] and are some of the very well-documented and widely cited research papers on smart transportation system. The most widely used is the VANET, relevant details presented in the following paragraphs.

A VANET is an innovation wherein vehicles go about as nodes to build a mobile ad-hoc network.

In a VANET, there are three sorts of communications viz. a vehicle to (vehicle-vehicle correspondences- V), vehicle to Road Side Unit (RSU) and RSU to RSU; the correspondence framework as shown in the Figure 11.3.

In VANET, vehicles communicate with RSU and different vehicles remotely. So they can, without much of a stretch, get data about street traffic, congestion or work out and about, street accident and so on. By utilizing this data, drivers can get ready to make intelligent decisions as per street conditions like changing the routes to the destination, unnecessary slowing down their speed, and so forth. They can likewise get some information about course to objective, accessibility of parking area, hotels, gas station, clinics in an obscure area [13].

Although a VANET is a kind of MANET it varies from MANET in the following manner.

Figure 11.3 Architecture of vehicular ad-hoc network.

- A VANET is described by quickly changing yet to some degree unsurprising geography.
- Network geography changes quick so fracture routinely happens.
- Due to rapid of hubs distance across of organization in VANET is generally little.
- Redundancy is restricted both transiently and practically.
- Do not require any framework.
- Predictable topology (utilizing advanced guide).
- No issue with power [14, 15].

VANET is also increasingly used in autonomous driving environment in addition to the predictable topologies covered above. Review of the same is taken briefly in the following paragraph.

11.4.1 Use of VANET in Autonomous Driving

Since long, many research groups on VANETs and autonomous vehicles dealt with various solutions for vehicle safety and automation, respectively. People were generally stressed over potential road safety, including the chance of their vehicle being hacked, accidents, or severe injuries because of the modification of safety messages. Notwithstanding, there is no rejecting that the world frantically needs options in contrast to human drivers. People make numerous careless mistakes that end up guaranteeing a considerable number of lives on a yearly premise. Less truly, helpless choices out and about typically add to traffic congestion.

   Real-Time Guarantees: Autonomous driving applications, VANET are utilized for crash evasion, danger cautioning and safety warning messages, so applications require time boundaries for message reception.

Smart Parking: An incorporated system of autonomous vehicles could incorporate self-driving taxicabs and independent vehicle sharing. A system of independent vehicles could make it practical to present savvy turnpike paths, on which the vehicles move in detachments to build throughput of the streets. Smart parking frameworks could likewise be actualized, whereby driverless vehicles drop their travelers off, get a parking spot themselves and park near one another. This spares space while conceivably delivering stopping offenses. Different applications may incorporate autonomous commercial vehicles that utilize at night to enhance street space. This would reduce human resources and traffic jams and accomplish future planning to build devoted highway paths for autonomous vehicles [16].

The term VANET alludes to endeavors to collect, store, and give ongoing traffic data to augment the use proficiency, give the good safe vehicles, and reduce energy consumption by applying propelled gadgets, data, and media transmission advancements into streets, cars, and goods.

VANET can fundamentally add to a cleaner, more secure, and more proficient vehicle framework in autonomous driving. Subsequently, VANET has become the focal point of various approach and authoritative activities in Europe. Furthermore, the European Commission has set out the legal system to quicken these imaginative vehicle advances across Europe.

However, individuals also gave loads of positive reactions to utilizing self-driving vehicles, such as better transportation for the older and fewer traffic jams or accidents. So, people are worried about self-driving vehicles being safe.

11.4.2 Security Requirement in VANET

Security is the primary concern in VANET in terms of cybersecurity component and physical road safety with the lack of central authority that can identify the legitimacy of the participant.

Ensuring the VANET administrations against the assaults many security natives ought to be thought of. Security requirements in vehicles are contrasted from the basic security needs, as shown in Figure 11.4. The salient points in this regards are as follows:

- Verification or authentication: The validation procedure needs to confirm the driver's identification or a person's identification approved to drive the vehicle.
- Integrity: Sensitive data and information obtained from different data resources like RSU, sensors, OBU, V2V, V2I, V2X ought not to be either revised or adjusted by an attacker.
- Non-repudiation: Once data transmission is done, then communicators cannot deny that data is sent by the sender and received by the receiver.
- Confidentiality: The privacy of transmitted data must be guaranteed in all vehicle frameworks (RSU, OBU, GPS, sensors). Data should not reveal in front of a malicious attacker.
- Access control: The vehicle frameworks are a vehicle that implies that has a place with people or to lawful substances. An ever-increasing number of secure methods must carefully constrain access to frameworks.
- Security and privacy: Vehicular frameworks are personal gadgets.

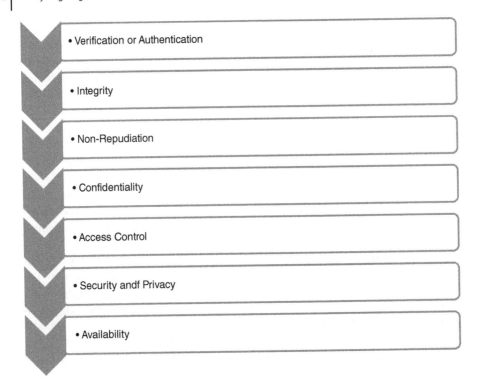

- Verification or Authentication
- Integrity
- Non-Repudiation
- Confidentiality
- Access Control
- Security andf Privacy
- Availability

Figure 11.4 Security requirement in VANET.

- Their tasks are administered by practices and guidelines. Therefore, the protection of the driver and travelers is of very high significance as one of the significant security angles.
- Availability: The vehicular frameworks have a long lifetime. The best possible working on autonomous vehicular subsystem Figure 11.4. Security requirements in VANET are guaranteed throughout the vehicle's life [17].

In autonomous driving, responsibility enforcement is a significant issue; typically, drivers are fully responsible for their driving. Therefore, the data is obtained from vehicles used to distinguish or aid the obligation attribution. Generally, in accidents (insurance experts, police officers), specialists are qualified for the recognizable proof of duties.

The vast majority of individuals consider that privacy is a piece of security; also, if security is accomplished, at that point, we do not have to stress over privacy. However, be that as it may, there is an excellent differentiation between security and privacy.

Security implies accomplishing the "confidentiality, integrity, and availability," regularly known as the CIA triangle. Information security alludes to the information being put away safe from unapproved access, guaranteeing that information is legitimate and is consistently accessible at the point when required. In any case, then again, security alludes to the utilization of data. It is the essential fundamental freedom, or it very well may be communicated as instructive self-assurance; the capacity to choose what data about an individual can be gotten to and by whom. Regarding their relationship with each other, security measures are made to secure individual privacy. Security is the finishes of methods privacy. Location tracking: The exact area of the vehicle has become fundamental data for vehicle security, specifically against robbery and location tracking. Tending to the harmony between confirmation and non-repudiation versus protection is a significant challenge for the vehicle's safety plan in autonomous driving. The major challenge is the trade-off between anonymity and authentication.

In VANET, nodes need interesting security prerequisites to accomplish the security required for safety. This is because there are so many application areas in which blockchain provides security to make smooth functioning and create comfort zones like voting, agriculture, banking, online music system, land registry, autonomous driving, and VANET, which is deliberated in this chapter.

11.5 Blockchain Provide Anti-Counterfeit Solution to VANET

In order to accelerate the performance of VANET and overcome the security loopholes of intelligent transport, VANET comes into a new era, namely blockchain technology.

VANET is now experiencing tremendous growth in autonomous vehicle technology. However, it is still more prone to various security vulnerabilities. This is because many traditional security techniques are incapable of providing high security for message sharing. The ultimate aim of data transmission in VANET is that data should be reached at destination securely without modification of contents. Recently, blockchain technology provided security and reliability in peer-to-peer, mobile adhoc network (MANET) with similar technologies like VANET. However, the components and entities in VANET susceptible to faults and becomes malicious attacker's target, they have difficulty trusting blindly on other vehicles in the network.

In VANET, roadside units (RSU) units act as a trusted third party, but if they compromised, it results in serious security and privacy challenges [17]. Safety messages transmission among vehicles and RSU will require the fast processing of safety messages within a fraction of a second. Traditionally, third parties are responsible for maintaining and executing contracts, but the blockchain concept aims to replace transactions between them. A smart contract is intended to guarantee both parties that the other party will keep its promises with certainty. Smart contracts can thus overcome moral risks such as strategic insolvency and drastically reduce the costs of review and enforcement. Smart contracts allow intelligent contracts to be created without the need for a third party or even third party use. One of the main objectives of blockchain and intelligent contract technology is to create fully automated intelligent contracts without human involvement [18].

In VANET, security messages and event messages should be dispersed safely because it legitimately manages complex issues like severe injuries and even demise.

Be that as it may, as its MANET, there are many odds of pernicious nodes sent modified or fake data or, once in a while, they alter messages securely. Moreover, some intruders will not transfer or even deliberately alter the necessary security messages in specific situations before sending to the receiver, bringing about longer delay or fatalities. Other than this, qualities of VANET (e.g., high portability, instability) which are unmistakable from different remote correspondence networks, have made VANET helpless to various inner and outer attackers [19].

In this paper, the authors [20] proposed blockchain to provide anonymous authentication and privacy preservation by providing pseudonym ID. RSU acts as miners to validate the vehicle's legitimacy by using a consensus mechanism. Moreover, the Merkle hash tree used to understand authentication records in real time. Timestamp field in header creates resistance against replay and Man-in–Middle attack. By varying the value of nonce in the session, the key achieves non-traceably. As blockchain is immutable because each block stores the hash value of its preceding block, no one can easily modify the message and preserve integrity.

To handle the distributed data VANET generates, the author used consortium blockchain for dynamic key updation and efficient group key distribution in the paper [12]. Furthermore, the authors proposed a certificate-less authentication scheme for edge computing assisted VANET that provide security to data communication between vehicle-RSU.

Furthermore, tragically, some selfish and avaricious node may not help other people; however, they will take their benefit. To adapt to this challenge, incentive mechanism is commonly considered as promising arrangement. In this paper [21], the authors plan a Bitcoin-based secure and dependable incentive scheme for helpful vehicular delay-tolerant networking. Bitcoin is the notable overall digital currency and computerized installment framework whose usage depends on cryptographic strategies, making it conceivable to build up a down to Earth credit-based incentive scheme concerning the vehicular organizations effortlessly.

The author considered scope only within the country in the paper [22]. The proposed method is scalable and would deduct latency because of its distributed nature and vehicles considered as having highly computational power. In the proposed method, authors used public blockchain to handle event messages such as traffic jam, accident, etc. When the vehicle receives an event message from another vehicle, the receiver first checks the sender's trust level by collecting information of the sender from its neighbors. If the sender is trustworthy, then and only then it keeps in the local memory pool; otherwise, the message is discarded. Then, the event message is broadcasted and validated by miners in the network by computing parameters like sender vehicle's previous trust level, proof of location, the freshness of the message by checking the timestamp. If the received message is valid, then the senders trust level increases. Miners places block permanently in blockchain by using proof of work mechanism. This mechanism prevents the network from denial of service attack, spamming, and injecting false information.

In the next era of a VANET, all vehicles will be connected through an ad-hoc network or the Internet. Blockchain technology is predicted to provide tremendous security and support for vehicle safety messages at an affordable cost [23]. It is an effective distributed ledger that can solve the security threats suffered by the ad-hoc nature of the vehicle. Although different investigations have explored the incorporation of blockchain technology with VANET, many aspects still unexplored. Based on introducing the basic principles of VANET and blockchain, thus conducted extensive discussions on existing blockchain application from a security perspective, with particular emphasis on integrating blockchain technology with VANET. The categories and explanations of various solutions proposed in recent research include security, trust management, certificate management, and privacy protection in VANET. Based on this survey, some open topics and future research directions are explored.

After discussing the blockchain, VANET, and security implications in light of smart transport, it is now worthwhile to review the emerging directions.

11.6 Blockchain Based VANET: Emerging Directions

11.6.1 Toward Fulfilling Sustainable Development Goals (SDGs)

As already revealed, the upcoming trend is to adapt to blockchain technology all over the globe. This is owing to its innovative properties. Furthermore, embedding with the transport system and smart city environment provides a win-win scenario to attain the SDGs. Many exciting projects are in progress in this direction. Under the coordination of Soumaya Ben Dhaou, the Unite for Smart and Sustainable Cities (U4SSC) initiative, a research coordinator and other experts from several organizations on blockchain and smart and sustainable cities, are in a process to examine the current progress and applications of blockchain technology in these areas in various verticals of smart cities and intelligent transportation. The said initiative was created to make decisive technological efforts to make cities more sustainable and to address several objectives set out in the SDGs, including goal 11 for sustainable cities and communities and the creation of smart cities. Efforts in this direction are expected to facilitate the development of initiatives that use blockchain to support cities, rural areas, and communities facing economic, social, and environmental challenges in line with the SDGs [24, 25]. In this context, the foremost reason, Hyperledger blockchain chose smart cities is its ability to interact with consortia. Of the 17 SDGs, four are of great interest to blockchain researchers—good health, better and sustainable cities and communities, clean water and sanitation and affordable and clean energy. Sensible solutions for smart city platforms include IoT, big data, the Internet, and energy that corresponds to a conducive environment for smart transportation are coming up rapidly [26]. The novelty of the approach is to design a decentralized platform that combines IoT and blockchain technologies for intelligent transport systems. The main characteristics of blockchain technology are decentralization, transparency and immutability, and the biggest advantage of the use of blockchain is trust in public institutions. The decentralized platform combining IoT and blockchain technology with intelligent transport systems will help to improve existing transport systems and overcome data security problems through transparency and trust. Even there are reported instances of decentralized data management systems for smart and secure transport that uses blockchain and the IoT to create a sustainable smart urban environment that solves data vulnerability problems. Blockchain's trust, openness, and transparency are built into its design, and its benefits extend to impacts on multiple Sustainable Development Goals (SDGs).

11.6.2 AI-Powered Blockchain for VANET

Today, blockchain technology, the IoT and AI are recognized as innovations with the potential to improve current business processes, create new business models and disrupt entire industries. In other words, smart contracts are an essential link between the three building blocks of IoT, blockchain technology, and VANET. Until now, the focus has been on linking blockchain technology to one or the other of these innovative technologies, such as IoT and AI, but not on applying all three innovations. For example, blockchain technology is used to authenticate IoT network subscribers, increase trust and manage the identity of IoT devices. In projects based on AI, blockchain technology can create decentralized and transparent networks that can be accessed from anywhere globally in a public blockchain network situation. Therefore, they have become popular in VANET applications.

There are growing applications on the theme of AI-enabled blockchain application for VANET. For example, an accident detection and validation system for detecting accidents by sending data to the Alicia Applied Intelligence Blockchain by performing MNS is recently reported. In VANET, artificial neural networks (ANNs) are embedded to select nodes to be excluded from the consensus process. To solve this problem, VANET empowered with an ANN to automatically select nodes that are excluded from consensus processes is employed [27].

11.7 Conclusion and Way Forward

The combination of blockchain technology and AI is still an unexplored area. Simply put, it has the potential to use data in ways that unimaginable before. Technology itself has received its fair share of scientific attention. However, the projects devoted to it are scarce, and from this viewpoint, immense research potential exists for applying above said technologies in the VANET domain. As reviewed in this chapter, the true potential of AI, the IoT, rather more specifically Industrial Internet of Things (IIoT) and 5G technologies can synergize with the intelligent transport system in general VANET in specific. If these technologies are integrated into the blockchain, smart and powerful transport systems with ad-hoc networks for vehicles can be realized. IoT technology empowered by AI, Neural Networks and Data Analytics can enable intelligent and optimized vehicles, computers, and people.

As blockchain technology can be used to authenticate IoT network subscribers, an increase in trust and management capability is evidenced by many researchers. This will be a boon for the VANET. Furthermore, blockchain acts as an information exchange medium between devices and IoT centers and databases and ensures a security-based framework for data exchange between vehicle-based MCs on the network. In projects based on AI, blockchain technology creates a decentralized and transparent network that can be accessed from anywhere in the world in a public blockchain network situation a scene emanating favoring the integration of these advanced technologies [28].

As already mentioned in this chapter, blockchain and AI technologies are based on the immutability of the data they collect. Therefore, the VANET scenario will be more beneficial if all the data and variables flowing into the system are subject to decision making through machine learning. The benefits of blockchain technology, including an invariable record of the data, variables and processes empowered by the AI in its decision-making process, is all set to revitalize the entire scene. Moreover, blockchain technology can be used to speed up key public upgrades by automating steps based on the contracts. Therefore it has the potential to solve the authentication problems in VANET communication. The pattern recognition, computer-aided learning theory, AI, and machine learning, i.e., the use of statistical techniques to give computer systems the ability to improve the performance of a particular task or data have the best chance of bringing about beneficial changes to ensure full information security in VANET. Blockchain technology avoids trusting third parties, thereby protecting against individual fault points and other problems. In other words, the smart contract is the most crucial link between the three components of the IoT and blockchain technology. Furthermore, using federated learning methods, approved supernode blockchains are maintained that implement computing and storage resources, store federated data models, learn raw data from decentralized parties, track the use of data for further testing, and ensure data security and privacy, all these capabilities are far useful in the VANET vertical.

In summary, blockchain technology improves data management in IoT devices due to its transparency, trustworthiness, truthfulness, immutability, security, and privacy. As a distributed registry, the blockchain prevents data leaks and single point failures caused by authorized administrators, and their anonymity to a certain extent protects the privacy of data owners. Therefore, the use of blockchain technology can improve privacy in-vehicle IoT systems. The scholarly literature indicates this trend of the combination of blockchain technology, IoT devices and AI that can unleash new business models for monetizing IoT devices [29, 30].

One of the leading open problems in this area is in Unmanned Aerial Vehicles (UAVs). UAVs have many data security and privacy concerns, and researchers worldwide have developed many solutions to protect data from cyber-attacks. Although some researchers have proposed blockchain solutions, their solutions suffer from high costs for data storage and network latency, reliability, and bandwidth. Some researchers have proposed solving the above problems with an interplanetary file system (BC) based on a secure drone communication scheme over a 6G network [33].

In summary, blockchain technology can be used in the transport sector to build intelligent transport management systems. The core of the problem lies in the timeliness, accuracy, and efficiency of the data, and it can solve this problem perfectly. Moreover, the government promotes blockchain technology to effectively improve intelligent transport systems and enhance transport in a more orderly manner. The distributed ledger capability can support a programmable economy since Bitcoin, and other cryptocurrencies already depend on it. Some derivatives like wind passé, blockchain technology promises to be the missing link that allows peer-to-peer contracts and behavior without a third party having to certify IoT transactions. IoT data and cars are not far away and will implement various blockchain applications in the coming years. They will soon use blockchain to link tolls and implement smart contracts across a wide range of industries in the coming months.

From the SDG perspective, the government heavily promoted blockchain technology and could lead to several smart cities that would promote sustainable urban development. In addition, blockchain and smart transport will connect thousands of homes and offer the public a more efficient and cost-effective way of life and access to health care and education.

The future is all about the empowered data-driven networks (DDNs), which is all set to provide smarter services, better quality of experience (QoE) for local users, safer data storage, sharing and analysis, better user privacy protection, robust and trustworthy network management, decentralized routing and resource management for VANETs. Despite the challenges outlined in this chapter, the blockchain-based IIoT, empowered by AI for IOVs, will provide safer vehicle environments.

Bibliography

1 Zwitter, A. and Hazenberg, J. (2020). Decentralized network governance: blockchain technology and the future of regulation. *Frontiers in Blockchain* 3.

2 Düdder, B., Fomin, V., Gürpinar, T., Henke, M., Iqbal, M., Janavičienė, V., Matulevičius, R., Straub, N., and Wu, H. (2021). Interdisciplinary blockchain education: utilizing blockchain technology from various perspectives. *Frontiers in Blockchain* 3.

3 Xiong, H., Dalhaus, T., Wang, P., and Huang, J. (2020). Blockchain technology for agriculture: applications and rationale. *Frontiers in Blockchain* 3.

4 Leible, S., Schlager, S., Schubotz, M., and Gipp, B. (2019). A review on blockchain technology and blockchain projects fostering open science. *Frontiers in Blockchain* 2.

5 Wang, Q. and Su, M. (2020). Integrating blockchain technology into the energy sector – from theory of blockchain to research and application of energy blockchain. *Computer Science Review* 37: 100275.

6 Sengupta, U. and Kim, H. (2021). Meeting changing customer requirements in food and agriculture through the application of blockchain technology. *Frontiers in Blockchain* 4.

7 Rejeb, A., Keogh, J., and Treiblmaier, H. (2020). How blockchain technology can benefit marketing: six pending research areas. *Frontiers in Blockchain* 3.

8 Andreev, R., Andreeva, P., Krotov, L., and Krotova, E. (2018). Review of blockchain technology: types of blockchain and their application. *Intellekt. Sist. Proizv.* 16 (1): 11.

9 Sabry, S., Kaittan, N., and Majeed, I. (2019). The road to the blockchain technology: concept and types. *Periodicals of Engineering and Natural Sciences (PEN)* 7 (4): 1821.

10 Anon. (2020). Smart and secured transportation system using IoT. *International Journal of Innovative Technology and Exploring Engineering* 9 (6): 95–99.

11 Abu-Lebdeh, G. (2017). Smart transport systems: planning and designing transport systems to support public health. *Journal of Transport & Health* 5: S34.

12 Anon. (2020). Guest editorial: intelligent transportation systems in smart cities for sustainable environments (SCfSE). *IET Intelligent Transport Systems* 14 (11): 1351–1352.

13 Gómez Mármol, F. and Martínez Pérez, G. (2012). TRIP, a trust and reputation infrastructure-based proposal for vehicular ad hoc networks. *Journal of Network and Computer Applications* 35 (3): 934–941.

14 Gadkari, M. (2012). VANET: routing protocols, security issues and simulation tools. *IOSR Journal of Computer Engineering* 3 (3): 28–38.

15 Balon, N. and Guo, J. (2006). Increasing broadcast reliability in vehicular ad hoc networks. *VANET '06: Proceedings of the 3rd International Workshop on Vehicular Ad Hoc Networks*, 104–105.

16 Anon. (2021). LTA | industry & innovations | industry transformation map | academies. [online] Lta.gov.sg. http://www.lta.gov.sg/ltaacademy/doc/J14Nov_p05Tan_AVnextStepSingapore.pdf (accessed 27 May 2021).

17 Sheikh, M.S., Liang, J., and Wang, W. (2019). A survey of security services, attacks, and applications for vehicular ad hoc networks (vanets). *Sensors* 19 (16): 3589.

18 Liu, H., Zhang, Y., and Yang, T. (2018). Blockchain-enabled security in electric vehicles cloud and edge computing. *IEEE Network* 32 (3): 78–83.

19 Broby, D. and Paul, G. (2017). Blockchain and its use in financial settlements and transactions. The Journal of the Chartered Institute for Securities and Investment (Review of Financial Markets) [online]. https://strathprints.strath.ac.uk/59818 (accessed 27 May 2021).

20 Engoulou, R., Bellaïche, M., Pierre, S., and Quintero, A. (2014). VANET security surveys. *Computer Communications* 44: 1–13.

21 Maria, A., Pandi, V., Lazarus, J., Karuppiah, M., and Christo, M. (2021). BBAAS: blockchain-based anonymous authentication scheme for providing secure communication in VANETs. *Security and Communication Networks* 2021: 1–11.

22 Li, G. and You, L. (2021). A consortium blockchain wallet scheme based on dual-threshold key sharing. *Symmetry* 13 (8): 1444.

23 Park, Y., Sur, C., and Rhee, K. (2018). A secure incentive scheme for vehicular delay tolerant networks using cryptocurrency. *Security and Communication Networks* 2018: 1–13.

24 Shrestha, R., Bajracharya, R., Shrestha, A.P., and Nam, S.Y. (2020). A new type of blockchain for secure message exchange in VANET. *Digital Communications and Networks* 6 (2): 177–186.

25 Koduri, R., Nandyala, S., and Manalikandy, M. (2020). Secure vehicular communication using blockchain technology (No. 2020-01-0722). SAE Technical Paper.

26 Ben Dhaou, S. and Backhouse, J. (2021). Blockchain for smart sustainable cities. [online] Collections.unu.edu. http://collections.unu.edu/view/UNU:7878 (accessed 27 May 2021).

27 Anon. (2021). Blockchain for smart sustainable cities: report now available – operating unit on policy-driven electronic governance. [online] Egov.unu.edu. https://egov.unu.edu/news/news/report-blockchain-smart-sustainable-cities-2020.html (accessed 27 May 2021).

28 Anon. (2021). [online] https://www.frontiersin.org/research-topics/18154/blockchain-for-united-nations-sustainable-development-goals-sdgs (accessed 27 May 2021).

29 Maskey, S., Badsha, S., Sengupta, S., and Khalil, I. (2021). ALICIA: applied intelligence in blockchain based VANET: accident validation as a case study. *Information Processing & Management* 58 (3): 102508.

30 Anon. (2021). Blockchain and AI: a perfect match? | OpenMind. [online] OpenMind. https://www.bbvaopenmind.com/en/technology/artificial-intelligence/blockchain-and-ai-a-perfect-match (accessed May 27, 2021).

31 Peng, C., Wu, C., Gao, L., Zhang, J., Alvin Yau, K., and Ji, Y. (2020). Blockchain for vehicular internet of things: recent advances and open issues. *Sensors* 20 (18): 5079.

32 Li, X., Wang, Z., Leung, V., Ji, H., Liu, Y., and Zhang, H. (2021). Blockchain-empowered data-driven networks. *ACM Computing Surveys* 54 (3): 1–38.

33 Sreekantha, D.K., Kulkarni, R.V., and Gao, X.Z. (2021). Blockchain and 5G-enabled industrial internet of things: application-specific analysis. In: *Blockchain for 5G-Enabled IoT*, 531–569. Springer, Cham.

Part 5

Blockchain Technology and Cyber Security

12

Enterprise Blockchain

ICO Perspectives and Industry Use Cases

Ashish Seth[1], Kirti Seth[2] and Himanshu Gupta[3]

[1]*Inha University, South Korea*
[2]*Inha University in Tashkent, Uzbekistan*
[3]*Amity University, India*

12.1 Introduction

Blockchain is basically a record keep system that is decentralized, transparent, and immutable. Such features of this technology make it almost impossible for an intruder to do any sort of manipulation with the information stored in this system.

It is essentially a decentralized database also known as Distributed Ledger Technology (DLT) (see Figure 12.1) that is maintained across the entire network of nodes on the blockchain. It refers to chain of blocks which stores transactions, each block has fixed size and connects to succeeding block by a cryptographic hash value, therefore if any of the block got manipulated it would be immediately apparent that information has been tampered.

This chapter describes blockchain core principles and explores how the founding principles of blockchian can be used to develop real-life applications. It also covers the underlying blockchain mechanisms, the technical and adoption challenges, and the governmental and non-government use cases [1, 2]. Several application areas are a good fit for both permissionless [3] and permissioned blockchains [4]. The chapter focuses on the underlying technology and architecture supporting the enterprise blockchain platforms. It also examines does the platform architecture work well with the use cases that the developers want to enable? For example, there's a difference between some enterprise blockchains' focus and public blockchains.

The chapter is divided into two sections. Section 1, will cover details about token economics, the concept of Initial Coin Offerings (ICO) and enterprise blockchain. Section 2 will discuss various industry use cases to understand when to use blockchain. We'll look at some specific industries use cases and how blockchain fits in (or doesn't). And finally, we'll take a step back and look at the culture, regulations, and caveats surrounding enterprise blockchains. Reviewing the use cases and applications of blockchains from the technical perspective can help both technical developers to better understand how the technology could be improved and could also help decision makers to understand the pain points of the technology limitations and capabilities [5]. Let's develop a fundamental understanding of blockchain and enterprise use cases in general.

Emerging Computing Paradigms: Principles, Advances and Applications, First Edition.
Edited by Umang Singh, San Murugesan and Ashish Seth.
© 2022 John Wiley & Sons Ltd. Published 2022 by John Wiley & Sons Ltd.

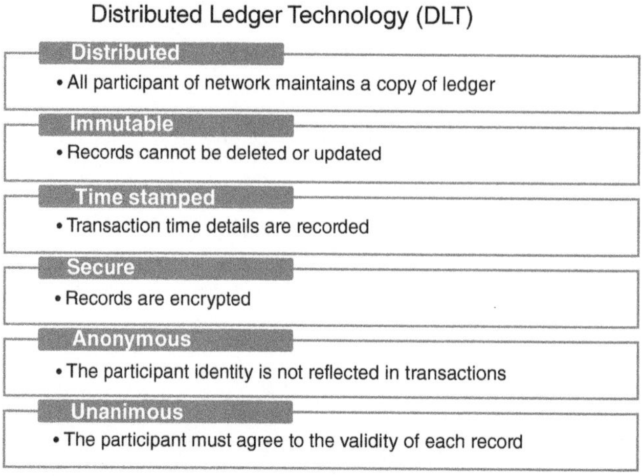

Distributed Ledger Technology (DLT)

Distributed
- All participant of network maintains a copy of ledger

Immutable
- Records cannot be deleted or updated

Time stamped
- Transaction time details are recorded

Secure
- Records are encrypted

Anonymous
- The participant identity is not reflected in transactions

Unanimous
- The participant must agree to the validity of each record

Figure 12.1 Distributed ledger technology.

Section 1

This section will cover details about token economics, the concept of ICO.

12.2   ICO Schemas and Culture

An ICO" is a novel, unregulated means of raising funds for a blockchain startup. It's novel because this model did not exist before blockchain and unregulated because blockchains lack the same strict governance that centralized institutions are often bound to. Companies developing a protocol often include a token within their platform. To distribute the token and raise money for their company, they will sell some of these tokens as seed funding. These tokens often represent some unit of value within their network. However, these tokens may be associated with the company in name only (see Figure 12.2).

Regardless, the concept behind these tokens is that their worth is tied to the company's value. As the company's value rises in value, so will the demand and, ideally, the tokens' value. Hence, the value can be defined by the community's amount of faith in the protocol's success. The greater the faith, the higher the value. This incentivizes the company to continue developing the protocol token's value goes up.

A few ICOs have happened in the past few years, many different projects, not limited to any particular industry. One example is Cosmos, a platform seeking to provide infrastructure for blockchain interoperability, raising 16,800,000 USD through their ICO. Another is Filecoin, aiming to create a distributed file storage network, raising 257 million USD.

12.2.1   Token Economics and ICO Innovation

There are two fundamental differences between IPO's and ICO's. ICOs are in general unregulated i.e., there is no government regulations or no authority oversee them.

ICO Structure

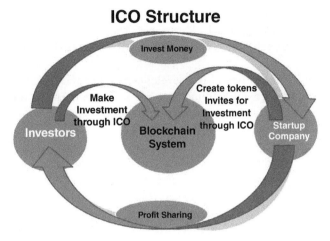

Figure 12.2 Initial coin offering schema.

Figure 12.3 Token economics.

Also due decentralized structure ICO's are more flexible than IPOs. ICOs are offered in various ways and exist is variable structures as shown in Figure 12.3.

If a company has a specific goal or looking for limited funding, then total token supply will be static and it offers ICO at a pre-set price. In other case, the company may have dynamic funding goal. In this case, token supply will be static but token distribution will depend on the fund received, i.e., the token price is directly proportional to fund received in ICO. Other scenario would be supply of dynamic token (i.e., tokens are not static). Here the price of tokens tend to static but supply of token depends on the funding received through ICO and the target of the company.

Let's take a deeper look at how ICOs innovate on previous fundraising mentalities. Previous fundraising activities, such as Venture Capitals, have often been through a select group of accredited investors. ICOs allow early adopters with a special interest and understanding in a particular blockchain use case to buy into the project early through the ICO, believing that it will come out on top. If the ICO's value is tied to the project's value, then those with high confidence in the project due to background knowledge will invest early. At least, that's the design; this allows project teams to get past an initial high capital requirement to build out their protocol while circumventing traditional, centralized fundraising models. This allows developers to get a cash infusion, build out a protocol, raise the token value, and then allow users to use them to request services.

Ideally, this would align the incentives of the development team and the early investors. Both want to see their token holdings increase in value, meaning the protocol needs to be developed as well. However, as we'll see, this is not the case today.

Let's take a look at how ICOs might incentivize open-source development. Open-source development is typically led by foundations, such as the Mozilla Foundation and Apache Foundation. These foundations are dependent on donations and volunteers, as the work is not funded as a company is. Hence, there is a dichotomy between the value created and the profit. While many individuals develop open-source software for the pure ability to benefit humanity, incentivizing open-source software development could make it accessible all around. The remaining majority is incentivized to make the software proprietary, causing technological lock-in. Companies hide trade secrets from each other to keep their leg up, slowing down intellectual growth.

ICOs now allow creators of open-source projects to monetize their efforts directly. Instead of looking for donations after pitching philanthropic individuals their vision, individuals can now create a digital representation of their platform's value. Additionally, as others build on top of that platform, the token will further increase in value.

12.2.2 ICO—Life Cycle

ICO life cycle can be defined using five stages,

1) Building a blockchain project
2) Identify an ICO model
3) Determine the regulatory and compliance laws
4) Writing smart contracts
5) Publicize project and promoting ICO

Well, let's start with the assumptions.

12.2.2.1 Building a Blockchain Project
Building a blockchain project requires some tokens. To make your case clear for your project, you'll need to write up a whitepaper to outline your project and how tokens come into play.

12.2.2.2 Identify an ICO Model
Next, we need to decide on an ICO model. This comprises of the following tasks

a) Fixed vs. dynamic supply
b) Stages of sale
c) Types of sale (Fixed price, reverse Dutch)

ICOs have sparked interest in the term "Tokenomics," referring to these digital tokens' economics. Because of space's nascence, several models are still being explored and understood, all focused on manipulating a few key variables and qualities. For example, a question that arises is whether to have a fixed or dynamic supply of tokens. A fixed supply may limit the number of people who can buy into the ICO, but a dynamic supply will make investors uncertain about the total number of tokens to be sold.

Next, there's the question about the number and types of stages to the ICO. Some projects have private fundraising rounds and pre-sales before the public sale. Additionally, there's the possibility of a dynamic price. A fixed price is straightforward: for some particular stage, sell at a set price. However, some ICOs aim to have dynamic pricing depending on the time and amount of tokens sold thus far.

For example, Gnosis, a smart contract prediction market project, used what's known as a reverse dutch auction. In this fixed supply and dynamic price model, the price starts incredibly high at first, slowly decreasing, meant to encourage investors to buy at the highest price, which seemed reasonable to them. However, the team underestimated the tokens' initial valuation: approximately 12 million dollars worth of tokens sold out within the first few minutes, accounting for only 5% of the supply. This led to Gnosis owning approximately 95% of the total token supply, leaving everyone confused. This ICO, which happened in early 2017, demonstrated how much more there was to learn about Tokenomics.

12.2.2.3 Determine the Regulatory and Compliance Laws

After deciding on the model, it's essential to determine the regulatory and compliance laws by which your model must abide and are currently unsatisfied. One question that might be asked is which countries legally allow for your sale of the token. Another is whether the type of token makes a difference in compliance. If seriously considering launching an ICO, hefty legal counsel must ensure that any regulatory concerns are handled.

12.2.2.4 Writing Smart Contracts

In the next stage, smart contracts need to be written. Every ICO is simply a smart contract or few which contain information about the ownership of every token, along with the functionality to send and receive tokens. The first step of development is to implement the functionality specified by your chosen ICO model, followed by inspecting the smart contract for security issues and vulnerabilities. Several smart contracts, which hold money, have been hacked in the last few years, such as the decentralized autonomous organization (DAO), leading to a great push for smarter contract security checks.

12.2.2.5 Publicize Project and Promoting ICO

Finally, publicize your upcoming project and ICO to potential investors to allow everyone to invest in your marvelous project. If that all works out, then you'll get funding for your revolutionary blockchain project! Legal uncertainty, as expected from so new a space, also permeates ICOs. Note that most of the legal precedents are still being established around the world.

12.3 ICO—Legal Uncertainties

Leaders in the space, such as Coin base, have written up material to serve as guidelines for evaluating tokens. Often, we find that ICOs are classified into two kinds of tokens: security tokens and access tokens. Securities are financial investment vehicles, such as stocks in a company. Take the DAO, for example. The Securities and Exchange Commission (SEC) ruled that all tokens distributed by The DAO are considered securities as they satisfied the security criteria. Recently, companies have hesitated to sell to US citizens to avoid crackdowns by the SEC.

On the other hand, access tokens are not meant to serve as any kind of financial investment. Instead, they simply serve to access services, such as how kickstarter donations provide the user with access to early or special goods from the team. Another caveat to consider is the reluctance or inability of project teams to keep their promise after an ICO. After all, after raising $200 million, what's the incentive to do anything else? There are no venture capitals (VC) or investors to demand money back—after all, the token does not provide such a privilege. Additionally, because most ICOs peak once and never recovers to that original value, early short-term adopters end up benefiting at the detriment of long-term adopters.

In 2017, investors in ICO pre-sales tended to sell right away, gaining four to five times returns on their investment instead of waiting for the long term. One proposed solution was for funds to be held in escrow, released only when the team hits particular milestones, to incentivize them to build out the project. Hence, teams are less likely to abandon their promises. However, then the problem of deciding who determines when milestones are paid out arises as well. This solution is not popularly used in ICOs.

12.3.1 ICos: A Double-Edged Sword

Recall, the purpose of an ICO is to remove regulatory bodies from the picture. However, without regulation, regular users are at higher risk. For example, thousands of investors bought into OneCoin, a cryptocurrency in name alone. This currency was purely run on SQL servers, entirely centralized. Everyone bought into it, assuming there was a blockchain use case underneath the project. As most investors are unused to projects springing up into the public eye without prior due diligence either by investment firms or by government bodies, they may fall prey to scams such as this.

Additionally, ICOs ironically result in centralization. It is noticed, founders and early investors end up disproportionately wealthier than others, frequently the result of "pump and dump" schemes, implying larger investors are artificially raising the price with the expectation of selling at a high value, leaving everyone else to bear the consequences. Finally, to demonstrate the success of scams in the ICO world, take the project known as MIROSKII, raising over 70 million dollars in funding without even having a real team! Unless Ryan Gosling is going under the pseudonym "Kevin Belanger" and night lighting as a graphic designer, he used his image without permission as one of the team members. Despite such a clear act of fraud, this project managed to fill its pockets obscenely. With the benefits of deregulation, such as larger freedom and accessibility to a larger set of investors, also come the consequences.

12.3.2 The Benefit of Traditional Venture Capital

The new paradigm of ICOs doesn't just have drawbacks for investors. Project teams also miss out on crucial benefits compared to VC funding. VC funding requires high levels of discipline. Investors seek out 10x, 100x companies; it's not that one can win the approval with just a lousy whitepaper. We'll need detailed financials to prove our merit and have answers to incredibly tricky questions about our company's fundamental assumptions. VCs do due diligence on projects both for the consumer and the founders: if a VC agrees to invest in a company, then that person has access to the VC's vast set of connections and experience; This can change the development of a project team. ICOs, on the other hand, are nothing more than lumps of money. They do not come with the support that a VC can provide. This model of funding comes with many further drawbacks.

12.4 ICOs: Drawback

As mentioned before, there is no central or trusted party to carry out due diligence on behalf of individual investors. Instead of VCs carrying out the groundwork and governmental regulatory bodies preventing poor investment decisions from reaching investors, ICOs are accessible to all; This is a huge distinction between ICOs and IPOs. Though IPOs require years of scrutiny and thought for even the most well-established companies, ICOs occur solely after a project idea and still raise exorbitant amounts of money. The burden of due diligence is now upon the investor. But how many investors read the whole whitepaper? How many investors look into every member of the team? Because of this, lots of ICOs unashamedly use marketing strategies to make the impression of profits for investors, even without any significant technical knowledge, to fulfill unrealistic expectations.

Additionally, investors also tend to be of a particular mindset: most early adopters already have a reasonably similar mindset, leading to a lack of diversity in thought with the project founders, possibly stunting growth. Finally, there is also a much larger number of investors to appeal to when launching an ICO, anywhere from hundreds to thousands, all of whom have different demands and expectations from the project. Thus, managing relations with all these investors can overwhelm small teams, a big reason why many companies prefer staying private, to focus only on the few large investors.

12.5 ICO–Statistics

Though ICOs provide tremendous access to funding, they also claim that: "be careful what you wish for." According to coinschedule.com, In 2017, we saw ICOs overtake VCs as the most popular form of funding for blockchain start-ups. By the end of 2017,

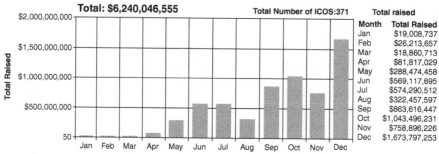

Month	Total Raised
Jan	$19,008,737
Feb	$26,213,657
Mar	$18,860,713
Apr	$81,817,029
May	$288,474,458
Jun	$569,117,695
Jul	$574,290,512
Aug	$322,457,597
Sep	$863,616,447
Oct	$1,043,496,231
Nov	$758,896,226
Dec	$1,673,797,253

Total raised are grouped by the ICO closing date and are valued using BTC exchange rate at the time. Data last updated on 26th September 2018 12:43 UTC

Figure 12.4 ICO funding 2017.

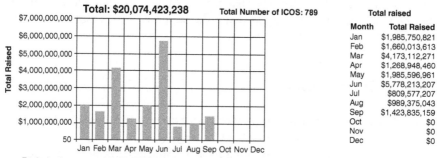

Month	Total Raised
Jan	$1,985,750,821
Feb	$1,660,013,613
Mar	$4,173,112,271
Apr	$1,268,948,460
May	$1,985,596,961
Jun	$5,778,213,207
Jul	$809,577,207
Aug	$989,375,043
Sep	$1,423,835,159
Oct	$0
Nov	$0
Dec	$0

Total raised are grouped by the ICO closing date and are valued using BTC exchange rate at the time. Data last updated on 26th September 2018 12:43 UTC

Figure 12.5 ICO funding 2018.

about 6 billion dollars was poured into blockchain start-ups through ICOs, over 25% raised in December alone (see Figure 12.4).

Compare this to approximately just over half a million raised through VC. In 2018, we saw even more raised through ICOs, totaling 20 billion dollars before the end of September (see Figure 12.5).

Instead, every third month seems to spike, followed by a couple of months of relatively less funding. VC funding has recently surpassed 4 billion dollars, up a great deal from 2017. Since ICOs seem not to provide long-term returns and have lost credibility as a funding mechanism, VC funding is starting to surge. You'll see in this slide the distribution of funding. As expected, the largest slice of the pie is infrastructure, at 25%. A majority of this comes from EOS, a distributed proof-of-stake blockchain. Infrastructure has a habit of having the most funding, given that the space is still new. Finance is second-largest, followed by communications, trading and investing, payments, and more.

Now that we have a clear working framework for categorizing and designing blockchain system architectures, we can start to see some of the patterns between these categories. Naturally, the community's best decision is to develop tools to make these common necessities easily available and secure for anyone else to build off of. As with any emerging technological field, the importance of infrastructure and accessibility for developers and companies looking to flesh out use cases is just as important as the technology itself. Reducing access barriers and uncertainty around a growing technology is crucial to capturing widespread adoption.

12.6 Blockchain Initiatives

One of the major challenges for blockchain adoption is the gap between the underlying technology and the understanding of the capabilities [3, 6]. There is a need to understand the existing and future state architecture before you begin selecting platforms and technology [7]. Many of the European Union countries took an initiative to supports blockchain projects across the European Union [8]. The principles objectives of most of these projects are to build a blockchain competence community and develop blockchain governance standards; the e-Estonia is a blockchain based program to supports number of services like e-identity, e-healthcare, and e-governance [9]. In 2019, the Filipino government decided to develop an Ethereum-based solution [10, 11]. The idea of *blockchain city* has already been promoted, this blockchain project aims to track booking services, passengers, tourist visas, and luggage [12]. South Korea's government declared to develop a blockchain-enabled virtual power plant in Busan city, the focus is to optimize power generation by integrating multiple energy resources [13].

12.7 Blockchain Technology Regulations Caveats

Careful handling of money requires several significant measures to prevent illegal activity. Money laundering refers to large sums of money between borders or between the underground and legitimate economy. Anti-money laundering (AML) laws aim to prevent these activities by ensuring that every financial intermediary is aware of the source's and destination's legitimacy. We may be able to tell why cryptocurrencies can cause issues. Blockchains circumvent centralized control and often contain protocols to obfuscate transaction and user information; many cryptocurrencies are currently at their height.

In the United States, both the SEC and the Financial Crimes Enforcement Network serve to enforce these laws, placing restrictions upon the types of activities that can be considered legitimate. The exchanges that are under compliance with AML include Coinbase, Kraken, and Bitstamp. A good number of exchanges are not compliant, given that a good amount of the blockchain community diverges from regulation as much as possible. Know your customer regulations are another kind of regulation imposed upon financial institutions to prevent knowingly or unknowingly enabling illegal behavior. For these reasons, KYC requires three things.

- First, these businesses must identify and authenticate clients; This is why we're required to submit great swaths of personal information to any bank with which we open an account; This allows banks to confirm that we are the same when we are transferring and claiming money. Additionally, it allows banks to associate any activity with a particular individual or entity when noticing red flags, possibly leading to further investigation.

- Second, they are required to evaluate the risk of a client. Each client may be holding and transferring to various entities different quantities of money, some of which may not be legitimate. It is the responsibility of the institution to determine beforehand whether the client might represent a risk.
- Finally, the institution must constantly watch for any indications of criminal activity. Often, a company will be forced to cut off business with a client who may be behaving suspiciously for protection's sake.

All these regulations are placed upon certain businesses to prevent any kind of financial circumvention around restrictions. However, because of the blockchain's decentralized nature, integrating these qualities into businesses is socially difficult, though exchanges like Coinbase will comply with these restrictions to serve audiences as large as possible. Entities that deal with money transfer services or payment instruments are known as money transmitters. To designate which entities are legally allowed to engage in such activities, there exists the Money Transmitter License issued by states within the US government. The process of achieving an MLT is sometimes known as a "financial colonoscopy" due to the depth of the application process. Some of the things that the New York Department of Financial Services has done before giving a license to a bank include but are not limited to auditing financial statements of the applicant business and any subsidiaries, investigating the personal financial records of directors, owners, and others, seeing a list of all lawsuits filed against any "control person" in the last 15 years, and performing third-party criminal and civil background checks.

This regulation depth is meant to protect consumers from businesses mishandling their money, but it makes performing these services enormously difficult. If that weren't enough, New York has a separate license exclusively for cryptocurrencies, that apply to anyone performing any one of five acts:

1) Receiving virtual currency for transmission or transmitting it;
2) Holding virtual currency for others;
3) Buying and selling virtual currency as a customer business;
4) Providing exchange services as a customer business; and
5) Controlling, administering, or issuing a virtual currency.

However, not all countries are as embracing cryptocurrencies and blockchain. Here are a few examples of countries pushing against cryptocurrencies and blockchain. In Bangladesh, it's claimed that a lack of regulation by a central bank makes cryptocurrencies dangerous. While they're not exactly wrong about the risk of cryptocurrencies to unknowing investors, they certainly punish beyond what many might say is a reasonable amount, threatening up to 12 years in prison for trading cryptocurrencies. In Bolivia, the central bank stated that it's illegal to use a non-government currency. In China, bans have been placed on practically all cryptocurrency and ICO-related activities. In Ecuador, restrictions have been placed on virtual currencies and, primarily, to protect the national digital currency, the first state-sponsored one in history. Iceland claims that purchasing and transferring digital currencies goes against the federal restriction against currency leaving the country, essentially banning cryptocurrencies. India shut down an exchange known as BTCXIndia, which was still deemed risky despite complying with AML/KYC regulations.

Section 2

This section will discuss various enterprise blockchain platforms and use cases they're geared toward—categorized primarily by their access type and consensus mechanisms.

12.8 Use Cases Blockchain Technology

In present time, blockchain potential has been identified by industries and it is realized that the blockchain framework can be used beyond cryptocurrencies (i.e., financial based transactions) Blockchain framework has been used to developed applications across the various industry such as healthcare, digital identity, supply chain, tourism, etc. (see Figure 12.6).

Blockchain systems are interesting to build and study; many business use cases don't require the same level of decentralization and restlessness. This is because public blockchains tend to assume nothing about users' motivations or incentives in the network, hence preparing them for the worst-case scenario. This ensures maximum restlessness, but the guarantee reduces efficiency greatly. On the other hand, many enterprise use cases require a lower level of trustlessness and decentralization in the system. Blockchain systems are not strictly decentralized but instead on a spectrum. Let's understand the different kinds of systems we can design. We'll first examine where enterprise blockchain systems fit in the whole space and the current platforms and other infrastructure available for use.

12.8.1 Blockchain Technology Use Case Auto Mobility

Mobility use cases combine blockchain, autonomous cars, and IoT. There's a non-profit consortium known as MOBI (Mobility Open Blockchain Initiative) to make

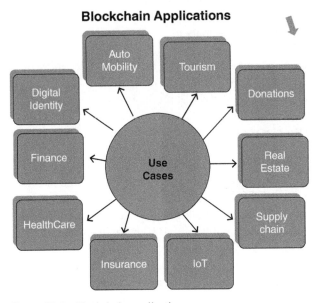

Figure 12.6 Blockchain applications.

mobility services efficient, green, affordable, safe, and free of congestion. Fun fact, MOBI was co-founded by a blockchain at Berkeley alumnus, Ashley Lannquist, doing a lot of exploration of the industry. One of their major use cases revolves around collecting a car's data, the obvious stuff like miles driven, MPG, and more microdata points like the force applied to the accelerator to inform safe driving cars. Will automatically compensate drivers for their data using token micropayments. Toyota and Jaguar/Land Rover are also exploring this concept. As far as other use cases, there's supply chain and provenance, so tracking car parts, or an immutable "Carfax"-style used car database. That's being done by a company known as carVertical. There are automatic machine-machine payments for electric vehicles and incentivizing autonomous vehicle decision making, like deciding when a car gets to merge.

There's car-sharing, like a decentralized Uber or Lyft, or a further decentralized Turo, Airbnb for cars. Some 25% of payments to either of these services go straight to the companies, not the people providing the cars.

12.8.2 Blockchain Technology Use Case Digital Identity

Companies focusing on decentralized digital identity are focusing heavily on a concept known as self-sovereignty of identity, which is the ability to limit the information you share in situations today where you might be asked for something like your passport, and also limit who has access to the information you do choose to share. If I want to be let into a building and am asked for my ID, I'm answering the question "Who are you?" when the only thing I need to provide is the answer to "Are you allowed to come in?" Self-sovereignty of identity tries to make that distinction clear by decoupling that information.

The Maltese government doing a pilot blockchain project to manage didgital identity. The key objective of the project is to improve the safety of personal information [14, 15]. Other company working on self-sovereign identity (SSI) is Vetri, formerly VALID. Vetri's model consists of a data wallet and a data marketplace. You can essentially reveal your data, therefore creating scarcity in the data market, and you'll be compensated in VLD tokens.

uPort focuses on creating a persistent digital identity that completely represents a person or organization that can make statements about who they are. It can be thought of as the ethereal equivalent of a Facebook profile in that you can log in to services with it representing your identity. uPort's app uploads your information to an independent decentralized storage platform and maintains its address on the Ethereum blockchain.

Sovrin's model is a lot like uPort's. Still, they custom-built a blockchain specifically for identity, which scales a lot better than the current iteration of Ethereum, potentially allowing it to be used on a global scale. Part of Sovrin's mission is not to deny identity to anyone for cost or accessibility. The key, of course, to all of these, and what makes decentralized solutions to identity useful, is that nobody should be in control of your identity but you.

Next, we're going to look at blockchain applications within the finance industry. You'll find that these use cases revolve primarily around tokenization, disintermediated value transfer, privacy, and traceability.

12.8.3 Blockchain Technology Use Case Finance

The finance industry is particularly well-situated for the blockchain revolution since many of the assets banks deal with are not necessarily physical, like stocks, bonds, etc. The Dharma protocol plans to move forward with this idea and release a token representing part of a debt asset. The concept is similar to securitization, where you turn an illiquid asset into a tradeable one. Many large banks are also just investing in cryptocurrencies, notably Goldman Sachs.

Blockchains and their associated assets also allow assets to pass across borders and jurisdictions largely unencumbered by regulation and intermediation. Many cryptocurrencies can be bought in one country and sold in another for local currency without paying a premium for exchange. This is particularly good news for large banks, whose large-scale transfers might have otherwise cost them a fortune. Interbank transfers represent one of the most high-profile and effective use cases of blockchain, and big names are paying attention—JP Morgan launched the Interbank Information Network, and Ripple has launched the Global Payments steering group, garnering support from groups such as MUFG, BAML, and the Royal Bank of Canada. SWIFT, the current go-to for global financial transfer, is also heavily invested in the space—their test project launched on Hyperledger Fabric and yielded positive results earlier this year.

Lastly, many finance giants are using blockchain to facilitate traceability or privacy. The Industrial and Commercial Bank of China uses blockchain as a verification mechanism for digital certificates, and Wells Fargo is using it to track securitized mortgages. JPM Chase has launched its Ethereum-Esque enterprise blockchain platform, Quorum, optimized for finance use cases—private transactions, permissioned network access, and smart contracts. Deloitte, KPMG, EY, and PwC are also banding together with Taiwanese banks to use blockchain to help audit financial reports. The transparency and immutability afforded to blockchain-based systems are particularly enticing for accounting use cases. Next, we'll be talking about blockchain applications in the healthcare industry.

12.8.4 Blockchain Technology Use Case Healthcare

In the healthcare industry, blockchain can provide utility by guaranteeing data persistence and availability. Ensuring the integrity and accessibility of medical records is extremely important, as it could save someone's life. However, this data should also be private. Medicalchain seeks to tackle this problem by storing access to medical data on the blockchain, only allowing access upon authorization from the user's mobile device. It's quite similar to a medical-focused uPort, except the data is stored in the same place it is now—in the hospitals you visit. Other groups using blockchains for medical data include MIT's MedRec, Taipei Medical University Hospital. Surprisingly, Wal-Mart, whose product will allow EMTs to view the medical record of unresponsive patients.

Blockchain is also particularly useful for providing financial incentives for good behavior, which, interestingly enough, could have a positive impact on the efficacy of modern healthcare. Sweatcoin is an app that will pay you to walk, tracking your GPS movements, and counting your steps using technology already built into your phone and rewarding you in Sweatcoins, which can be exchanged for things like PayPal cash, an iPhone, or product discounts. Sounds crazy, but Sweatcoin purportedly has "converted" some 2 trillion steps into cryptocurrency. Mint Health takes a similar but more medical approach, providing Vidamint tokens for good behaviors ranging from checking one's blood sugar to attending a health-related webinar to recording steps. People can exchange these tokens for rewards such as lower insurance premiums. Insurance as an industry is also slated to be radically changed by the blockchain.

12.8.5 Blockchain Technology Use Case Insurance

At the most basic level, insurance is something of a prediction market—you're placing bets based on the likelihood of some future outcome. This makes it a great industry to integrate with blockchain. Companies need to make decisions based on data they can trust in an industry with so much fraud and inefficiency. A blockchain can provide a very high level of security and transparency that would provide such trust. Accenture reports that 46 percent of insurers expect to integrate blockchain within two years. A blockchain solution would allow the insured party to log their claim and their evidence immutably on the chain and be validated by the network such that the insurer can take it as truth. Claims are a particularly dicey area of insurance—they take a long time to process and always involve two parties with asymmetric information at odds.

A robust blockchain solution to fix claims might involve the use of smart contracts that automatically dispense payments when a specific set of requirements for a claim are met. Aigang is exploring a solution like this—they are insuring smart devices and automatically processing claims by having them log their state on the blockchain. They currently support insurance for phone batteries and are working on implementations for smart cars, smart homes, and drones. Additionally, a group of European insurers has come together to form the Blockchain Insurance Industry Initiative (B3i), which intends to use blockchain to add security and transparency to current insurance services. Next, we're going to look at how blockchain interacts with IoT.

12.8.6 Blockchain Technology Use Case_IoT

IoT devices collect and push data, and blockchains verify and codify it. Many of the use cases we've discussed so far involve smart contracts making decisions based on parameters, such as for Aigang's smart device claim distributions. The data collection mechanisms are often IoT devices.

FairAccess is an access control framework for IoT that is based on blockchain technology [16].

The biggest benefits blockchains bring to IoT are security and trust. Blockchains improve IoT security and trust by forcing the IoT network to converge on truth rather than have a single, potentially compromised database providing false information for critical situations. Many current IoT systems have a single point of failure, which, when under attack, can result in catastrophic system failures like the Mirai Botnet attack in 2016. Groups such as the Trusted IoT Alliance aim to tackle these problems by using blockchain to set standards for what a "good" IoT node looks like and quarantine nodes that don't measure up while distributing the collected information on those devices to multiple sources. Next, we will talk about blockchain in the supply chain, which leverages IoT as well.

12.8.7 Blockchain Technology Use Case Supply Chain

Any enterprise has the possibility to win in the competition only when it forms the strategic alliance with the upstream and downstream enterprise [17]. The efficient application of supply chain management provides transparency and traceability to reduce risk by increasing awareness of cause-effect relations [18]. With the rise of supply chains, many massive corporations have attempted to incorporate blockchains into their business models. One example of this is Wal-Mart, which has decided to use a supply chain for provenance. Provenance is securing the traceability of specific objects by tracking their history.

This provenance is done through a blockchain, where products are traced through the companies they interact with, whether they are growers, distributors, or retailers. Along each step of the way, those handling the products will be required to create a transaction on the blockchain and sign them. With this technology, rather than recalling all potentially affected products when a subset of these products is contaminated, Walmart can use this supply chain to track where this illness came from and only remove the products that came from the same source, thus saving a lot of money, resources, and time.

Another massive corporation that has attempted to integrate the supply chain is Alibaba, the Chinese e-commerce conglomerate. Noticing that China has had an issue with counterfeit goods for decades, Alibaba, among other e-commerce corporations, has decided to mitigate this issue by using QR codes and RFID.

The blockchain solution has the potential to track accountability and monitor process flows between buyers and suppliers [19]. By tracking goods such as food, baby products, liquor, and luxury items, the likelihood of being fooled by counterfeit goods is decreased, while consumers are given more trust that their products are real. Despite this, there is still plenty of skepticism against using blockchain technology for provenance. For example, many say that using blockchain serves no better than simply using a centralized database, as the decentralized nature does not give any beneficial advantage over a normal database. Currently, Wal-Mart's blockchain information is all stored on IBM's servers, which essentially defeats the purpose of a decentralized system if all of the data is kept under a single central authority.

Another problem with supply chains includes the difficulty of tying physical objects to the digital world. For example, giving each bag of lettuce, a unique identification code printed on the bag can be easily forged or changed at any stage of its life. Walmart could use RFID tags on each of the lettuce bags, but the cost of this would far outweigh the potential gain, making it an unreasonable solution. Supply chains are unable to protect against many issues such as fraud that human inspectors can only detect. So while a blockchain would allow us to trust the information channel, the endpoints' input data to the block chain are still fallible. This endpoint verification problem, or, more informally, "garbage in-garbage out," is one of the largest barriers to large-scale blockchain adoption today.

12.8.8 Blockchain Technology Use Case Real Estate

Real estate is another industry being revolutionized by blockchain, primarily when it comes to land rights.

Blockchain land registry projects have cropped up all over the world, from Sweden to Ghana to India.

- Current systems rely largely on paper deeds that are extremely frustrating to keep track of and are often lost for good, especially in the wake of national disasters.
- Additional problems arise when we consider forged signatures, improper paperwork, and many other details that become important when one wants to prove ownership or change it.
- Lastly, corruption in governments and corporations today can interfere with property rights in many countries.

What good is a land title if you can't be confident that it is available, persistent, and valid? The blockchain solution is to create a hash of every land registry and store that hash on-chain. Only the person with the corresponding private key can claim ownership of the corresponding land title. The tamper-evident nature of cryptographic hashes and blockchains makes it impossible to change any part of a land title without alerting the entire system. Georgia and Sweden has implemented blockchain for land title registry and related property transactions [20–22].

Additionally, since the registry data is replicated in every node, accessing land titles cannot easily be blocked by a malicious attacker or a natural disaster. Lastly, a blockchain land registry could drastically decrease the amount of time it takes to transfer land. Sweden's Lantmäteriet, in conjunction with blockchain startup ChromaWay, said that their prototype cut a digital land registry's lag from 3–6 months to a few hours. Some other companies looking into this include Propy, whose pilot with a Vermont city was launched earlier this year. Zebi Data, an Indian blockchain startup partnering with Maharashtra's states Telangana.

12.8.9 Blockchain Technology Use Case Travel Tourism

Blockchain is also making waves in the travel industry, allowing people to keep track of their luggage, get better rates on extra rooms, participate in the sharing economy, and set up secure payment channels, all without paying an intermediary to handle it for them. Winding Tree uses blockchain to disintermediate the process of filling excess vacancies on flights and in hotels and has already partnered with Nordic Choice Hotels and Lufthansa, Swiss Air Eurowings. Bee nest is doing the same thing with home shares—it's like Airbnb, but without giving them a piece of the pie. Trippki is creating a decentralized rewards ecosystem with travelers and hotels, making rewards transferable while still allowing hotels to make specialized offers.

One of the largest problems facing the travel industry today is identity management. How can an airline be sure if the person getting on their flight is the person whose name is on the ticket? What happens if you lose your driver's license or passport before taking a flight? It turns out that this problem is challenging alone, but here's what companies in the travel industry are doing to solve it on their end. The IT company providing support to a lot of the airline industry has created the Digital Traveller Identity App that uses blockchain to link a traveler to their identity and allows airlines and agencies to easily verify their identity while keeping the privacy controls the hands of the user.

Additionally, Dubai's Immigration and Visa Department has partnered with a UK-based startup ObjectTech to develop a digital passport concept combining biometrics and blockchain technology. ObjectTech is expected to launch its pilot program at the Dubai Airport in 2020.

12.8.10 Blockchain Technology Use Case Foreign Aid

Blockchain can improve the efficiency and effectiveness of foreign aid. Today's aid distribution systems are fraught with problems—many intermediaries lie between the money you donate and the intended recipient, and each one takes a cut; plus, if the destination doesn't have adequate disbursement infrastructure, there's no guarantee anyone in need will see any of the money sent. Blockchain can help with this.

Foreign aid can be divided into cash and non-cash aid. Let's talk about non-cash aid first, though both benefit the same way from implementing blockchain technology. Non-cash aid systems that use blockchain include the United Nations World Food Program. Their refugee camps in Jordan currently distribute crypto-vouchers that refugees can exchange for food and other necessities.

For the most part, however, aid disbursement is heading in the direction of cash aid since it is more versatile and, thanks to blockchain, can now be traced from sender to recipient. Fraud and accountability problems are more severe for cash aid providers since vouchers like UNWFPs aren't necessarily going to be usable by any intermediaries. Implementing a blockchain and using it to transfer aid allows the sender to see where all of the money goes, and the immutable nature of the blockchain ensures that the transactions on-chain have not been altered. Consensys Social Impact's Project Bifröst goes as far as to cut all unnecessary intermediaries out altogether. Bifröst transfers aid as a stable coin (guaranteed to hold its value), which is to be exchanged for local currency via kiosks in the recipient country. Having only two intermediaries drops the transfer cost to as low as 1% of the transaction value, significantly lower than any transfer medium today. On the other hand, disbursements choose to maintain an intermediary network and demand strict accountability by logging each transaction on-chain. They exchange the transferred cryptocurrency at local banks. Next, we dive into generalizations that can make between use cases.

12.9 Blockchain Technology Use Case Generalization

Now that we've seen all these use cases let's finish off with some generalizations. These points are to understand and discuss any use case we come across or come up with. These points describe the properties of a good blockchain use case, along with the scenarios in which a blockchain is not needed or inferior to a centralized solution.

In general, discussion about generalizations can be done in smart contracts and applications of the blockchain. Here, we take use case generalizations in the context of enterprise blockchain use cases. Let's talk about the scenarios where a blockchain will work but is not necessary. We often hear the term "efficiency" in the context of blockchain use cases, but this isn't often applicable. For example, let's take Bitcoin. If

I'm buying coffee, it's less efficient to send Bitcoin and wait 10 minutes to confirm than to hand over a few dollars or use a credit card. However, it is more efficient to send value overseas with Bitcoin, which takes a mere 10 minutes, compared to the days that banks take to coordinate and the transaction fees they levy for international transfers of value. Hence, efficiency depends on context.

Additionally, data storage characteristics such as data immutability, integrity, suitability, and authenticity are possibly much lower costs without a blockchain. Redundant, mission-critical, fault-tolerant systems have been around for decades, and cryptography has been around for millennia. All of these properties have been solved individually. Blockchain simply ties them all together. Each of these bullets can be achieved using a subset of the technology that makes a blockchain. While blockchains will solve any of these individual requirements, they're over-engineered solutions to these problems.

As mentioned in several of the use cases, blockchains allow us to solve coordination failures. We can implement arbitrary incentive schemes, allowing us to create a system that incentivizes individuals to operate according to our expectations. Also, blockchain can be thought of as a "technological solution to a social problem." Theoretically, every blockchain protocol could be run by a single node. However, if only one person runs the protocol, we'll notice that we lose out on the guarantees of audibility and decentralized control. These properties are meaningful only in a social setting. When individuals don't trust each other, the blockchain allows them to coordinate between them without relying on some trusted third party. With this, blockchains can make commitments, fund public infrastructure, or do crowdfunding. The network will force the actors to honor their commitments, as was smart contracts' original intention.

Instead of bringing in lawyers to settle matters when things don't go according to plan, we're now able to rely on a smart contract to execute as intended, giving us the ability to believe in this code as law. Blockchains create a standardized platform for access and interaction. Because of this, we can combine the power of all users on a blockchain network to enhance everyone's capabilities. Given that all information in a blockchain is accessible to everyone, we can combine data silos between institutions. Any information collected or functionality provided by an app on a blockchain network is accessible to all other users on the blockchain, which can't be said about the Internet alone.

Also, blockchains enforce a common standard. As all users tap into the same protocol, they must also adapt to that protocol's specifications. Granted, that requires everyone to go through the trouble of adapting. However, once everyone's on the same platform, there are no longer format or syntax issues. Lastly, by combining resources and information from all parties, we can enhance everyone's user experience. Any app that exists on a smart contract platform has its data and functionality living in that platform. Any other app can leverage existing technologies on the same platform, creating a positive feedback loop and benefiting everyone. This is referred to as network effects, which is the increased value or potential of a product with every additional user. Like how more Facebook users make the platform more worthwhile for all users, more smart contract developers increase a platform's value. In this way, individuals are supporting the rest of the community while benefiting themselves as well.

Finally, the most abstract yet fundamental property of a good blockchain use case is pure decentralization; it means decentralization for the sake of keeping it out of the hands of a central authority. This is what Bitcoin aimed to do with banks. Although there was a central working solution, Bitcoin wanted decentralization nevertheless. In countries with significant amounts of distrust in central authorities due to corruption or inefficiency, as mentioned during the real estate section, blockchain might be useful. Blockchain provides a system for users to produce guarantees that a central solution cannot provide, such as censorship resistance and disintermediation of power. These properties are difficult to evaluate in terms of dollars and cents, but groups like cypherpunks and crypto-anarchists ask about finances second. For some individuals, self-governance and privacy are more important than any amount of revenue, making decentralization for decentralization's sake worthwhile. Perhaps the most astonishing property that a blockchain provides is globally recognized proofs.

Through blockchain, we can support globally recognized ownership, persisting across nations.

Now that we've finished talking about decentralized solutions' meaningful properties, it wouldn't be complete if we didn't go over the caveats. What are the costs of these properties of decentralization? What do we achieve better with centralized solutions? The overarching theme of centralized solutions is the benefit of independence. There's no need for consensus when a single party has the power to make decisions. Because of this, we get the following benefits:

- First and foremost, we have deep integration. A central solution has full control over everything under its umbrella. Apple is well known for taking advantage of this to control the user experience. When a blockchain attempts to upgrade its protocol, all users must upgrade or get left behind voluntarily. However, it's much easier to change individual components or entire projects with a central solution. Because of this, it's much easier for a central system to patch up bugs, such as security issues, than decentralized systems. A central solution does what it needs to do, unrestricted. Still, a decentralized system needs to reach a consensus with thousands of different actors to change anything at the protocol level.
- Another huge advantage of central solutions is efficiency. With centralized solutions, the cost of executing a program is about a million times less work than decentralized solutions; this is easy to see, as only one party is doing work, and it doesn't need to confirm the result of its work with anyone else.
- Also, only one store of data is required. The data don't be replicated across thousands of nodes.
- Also, access control is much simpler in a central solution, where it's much easier to restrict read and write permissions. In a decentralized solution with censorship resistance, we give up that control. Building off that, central solutions handle complexity well.

Centralized solutions have the advantage of handling messy situations with grace since you don't need every single entity to agree on every single outcome. If you integrate centralization with a blockchain solution, you lose out on most of the decentralization benefits. The main takeaway is that there are advantages to both centralized and decentralized solutions. Neither is universally better than the other; they each have their use cases.

12.10   Conclusion

ICOs allow early adopters with a special interest and understanding in a particular blockchain use case to buy into the project early through the ICO, believing that it will come out on top. This chapter focused on various industries worldwide to give you a sense of what is currently being pursued to see into big corporations' minds and get a sense of the patterns behind today's blockchain ventures. This chapter doesn't just illustrated a comprehensive list of blockchain use cases nor advocate for any particular industry or company. Instead, it provided a larger insight into some of the common use cases seen within the enterprise blockchain space.

However, the best solutions recognize when decentralization is critical to accomplishing some goal and don't get distracted when a blockchain is viable but doesn't make sense. A good blockchain use case is like an oasis in a desert. Mirages pop up all the time, but that doesn't make them the real deal. Be sure that you're able to justify why a blockchain works for your use case! Some use cases may be more appropriate for blockchain than others—the goal is to apply your knowledge of blockchain to these use cases to evaluate the use case quality.

The challenges that still needs to be addressed includes preserving full privacy, integrating with legacy system and ensuring compliance, and of course scalability, to address these concern will be main focus area of next era of blockchain as still this technology itself is evolving and lot of effort and experience is required to address them.

Bibliography

1 Correia, M., Veronese, G.S., Neves, N.F., and Verissimo, P. (2011). Byzantine consensus in asynchronous message-passing systems: a survey. *International Journal of Critical Computer-Based Systems* 2 (2): 141–161.
2 Wang, X., Zha, X., Ni, W., Liu, R. P., Guo, Y.J., Niu, X., and Zheng, K. (2019). Survey on blockchain for Internet of Things. *Computer Communications* 136: 10–29.
3 Nguyen, G.-T. and Kim, K. (2018). A survey about consensus algorithms used in blockchain. *Journal of Information Processing Systems* 14 (1): 101–128.
4 Cachin, C. and Vukolić, M. (2017). Blockchain consensus protocols in the wild. *Proceedings of the International Symposium on Distributed Computing (DISC'17)*. Article 1.
5 Clavin, J., Duan, S., Zhang, H., Janeja, V. P., Joshi, K. P., and Yesha, Y. (2020). Blockchains for government: use cases and challenges. *ACM Digital Library* 1 (3): Article 22. doi: 10.1145/3427097.
6 Clavin, J. and Duan, S. (2019). Global transformation with blockchain: from lab to app: workshop summary. https://carta.umbc.edu/workshops/workshopsblockchain-workshop2018, accessed May 13, 2021.
7 Seth, A., Aggarwal, H., and Singla, A. R. (2013). Framework for business values chain activities using SOA and cloud. *International Journal of Information Technology, Communications and Convergence* 2 (4). https://www.researchgate.net/publication/264823142_Framework_for_business_values_chain_activities_using_SOA_and_cloud.

8 Trustnodes. (2018). The European blockchain partnership signed, €300 million allocated to blockchain projects. Retrieved October 21, 2020 from https://www. trustnodes.com, accessed May 21, 2021.

9 e-Estonia. n.d. (2020). Home Page. https://e-estonia.com, accessed May 21, 2021.

10 Demirguc-Kunt, A., Klapper, L., Singer, D., Ansar, S., and Hess, J. (2018). *The Global Findex Database 2017: Measuring Financial Inclusion and the FinTech Revolution.* World Bank.

11 Zottel, S., Zia, B., and Khoury, F. (2016). *Enhancing Financial Capability and Inclusion in Sénégal: A Demand-Side Survey.* World Bank.

12 Asia Blockchain Review. (2019). Malaysia's Melaka straits city to become world's first blockchain city. Retrieved October 21, 2020 from https://www. asiablockchainreview.com/malaysias-melaka-straits-city-to-become-worlds-first-blockchain-city, accessed May 22, 2021.

13 Partz, H. (2018). Major South Korean city to build blockchain-enabled virtual power plant. Retrieved October 21, 2020 from https://cointelegraph.com/news/major-south-korean-city-to-build-blockchain-enabled-virtual-power-plant, accessed May 20, 2021.

14 Gräther, W., Kolvenbach, S., Ruland, R., Schütte, J., Torres, C., and Wendland, F. (2018). Blockchain for education: lifelong learning passport. *Proceedings of 2018 1st ERCIM Blockchain Workshop.*

15 Grech, A. and Camilleri, A. F. (2017). *Blockchain in Education.* European Commission.

16 Ouaddah, A., Abou Elkalam, A., and Ait Ouahman, A. (2016). Fairaccess: a new blockchain-based access control framework for the internet of things. *Security and Communication Networks* 9 (18): 5943–5964.

17 Seth, A., Agarwal, H., and Singla, A. (2012). International conference on advances in computer applications. Proceedings published by International Journal of Computer Applications (IJCA), Unified Modeling Language for Describing Business Value Chain Activities.

18 Frentrup, M., Theuvsen, L. et al. (2006). *Transparency in Supply Chains: Is Trust a Limiting Factor*, 65–74. Bonn: Trust and Risk in Business Networks, ILBPress.

19 Asharaf, S. and Adarsh, S. (2017). *Decentralized Computing Using Blockchain Technologies and Smart Contracts: Emerging Research and Opportunities: Emerging Research and Opportunities.* IGI Global.

20 Lemieux, V. L. (2017). Evaluating the use of blockchain in land transactions: an archival science perspective. *European Property Law Journal* 6 (3): 392–440.

21 Allessie, D., Sobolewski, M., and Vaccari, L. (2019). Blockchain for digital government: an assessment of pioneering implementations in public services. Technical Report. Joint Research Centre (Seville site).

22 Shang, Q. and Price, A. (2019). A blockchain-based land titling project in the Republic of Georgia: rebuilding public trust and lessons for future pilot projects. *Innovations: Technology, Governance, Globalization* 12 (3–4): 72–78.

13

Blockchain and Cryptocurrencies

Techniques, Applications, and Challenges

Snehlata Barde

Professor, MSIT, MATS University Raipur (CG), India

13.1   Introduction

The primary objective of security to make a shield for computers, networks, and applications that protect from those people who are not authorized and do not get permission to them for accessing or updating the content or complete information. After an evaluation of the new technologies blockchain provides, the facilities of the person that stored the transaction record digitally and the cryptocurrency is known as a product of blockchain.

13.2   Problems with the Current Transaction System

There are many problems that organizations, business companies, and share advisers are facing.They found that the current system has a number of challenges, some of which include

- Some additional charges have to be paid by the customers when they transfer money from one bank to another.
- The third party/mediator increase the transaction charge.
- Transaction amount size is fixed, whichmeans its range is defined previously.
- It is a prolonged and time consuming process.
- System fairness and transparency rates are low.
- Control and power in the hand of central authority.

13.3   Distributed System

To overcome the challenges of the current system and open the path for new technology development, a new system was invented, known as a distributed system.

- Decentralization of a central system for providing the facilities, all the nodes perform their work individually and complete the task with the coordination of each other and get the outcomes. The distributed system looks like a single logic platform where all the devices are connected in a network and sharing the data

Emerging Computing Paradigms: Principles, Advances and Applications, First Edition.
Edited by Umang Singh, San Murugesan and Ashish Seth.

that are stored on the Internet. It removes the command or control of the single party and gives an equal chance to all connected nodes.

- Designing and processing of distributed system is a challenging task. Every node has the responsibility to manage the task carefully. Every node is capable of sending and receiving the message to or from each other due to one node fault system have to tolerate this.

13.4 Cryptography

A technique where the sender can send the data or information, not in its original form, but converted into cipher text by the encryption and receiver get the original data or information by the decryption. Cryptography used both symmetric and Asymmetric encryption technique for converting the message using the public and private key. Symmetric encryption used the same public key for encryption at the sender end and decryption at the receiver end, but, in asymmetric encryption used two key, public and private, one for encryption and one for decryption; secure transmission of the message is indicated in Figure 13.1.

- Two keys, public and private, are related.
- The private key is secure. It is not shared with anyone, but the public key is shared in the network.
- A public key is derived by the private key but not vice versa.
- One way function is used to generate a public key from a private key.

13.5 Cryptocurrency

Cryptocurrency is a new word to explore nowadays; it is defined as the simplest form like virtual money, it kept the account or wallet in the form of crypto that refers to processing and creation of digital currency and focused on transaction of the system that is decentralized. Its design, modification, and control free from the government [1].

Cryptocurrency is a combination of some conditions, such as it does not require control of central authority; it is comfortable in the distributed system. After creating a new cryptocurrency, the circumstances of its origin and ownership are defined previously, and ownership is proved by the cryptographically the system allows the

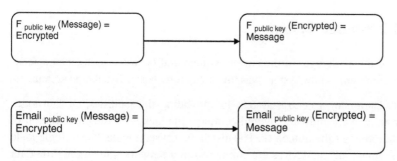

Figure 13.1 Secure transmission of a message.

permission of transaction that can change the ownership of units. A digital currency or cryptocurrency exchange provides the facility to the customer they can trade for assets using digital or cryptocurrency.

13.6 Digital Money

Digital money is not similar to traditional money like cash, coin, or dollar. It is totally in electronic form and can be easily transferred or exchange by the online technique, smart card, credit card, and smartphone. It reduced the physical payment mode and made an easy transaction for all anywhere and anytime. Nowadays, a form of digital money that is widely used is known as Bitcoin, Ripple, and Etherum. They are comfortable within the blockchain network. It is more secure and safe [2].

13.6.1 Bitcoin

A new digital or cryptocurrency Bitcoin is not a tangible coin created in 2009 by Satoshi Nakamoto. It is more popular and used BTC as an abbreviation. Bitcoin works under the control of the decentralized authority. Bitcoin code's run between the collections of nodes and store it in the form of blockchain in the Bitcoin system. Its transaction fee is less than the traditional mechanism and used peer-to-peer method for instant payments. Physical payment is not allowed; only balance is store in the ledger that is transparent for everyone [3].

13.6.2 Ripple

The system that settles the gross amount in real time is known as ripple. In 2012 US-based a company invented a ripple lab that creates the network for the currency exchange. It is an online secure payment system that works globally. Ripple is also working on the distributed system and supports the cryptocurrency. Ripple relies on a common shared ledger that stores all the information about the account. Ripple sends the immediate notification after the transmission of payment and indicates the balance amount for the valid account.

13.7 Blockchain

Nowadays, people want to de reduce human activities, and they want to do work through the machine automatically; for this, they made software to perform this type of task such as online banking, online ticket booking, etc.

"A blockchain is a collection of continuously growing records; these records are called blocks, linked together and used as cryptography for security purposes." A new technology that provides the solution of record keeping and makes a possible huge transaction in the future is known as blockchain technology introduced by the Stuart Haber and W. Scott Stornetta in between the 80s to the 90s. A blockchain is nothing but a collection of records, which is also called the block. Figure 13.2 indicates the blockchain structure that contains data, pre-hash, and Hash value [4].

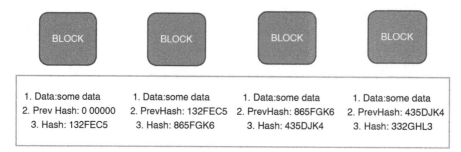

Figure 13.2 Blockchain.

13.8 Structure of Block

Blocks are connected to each other and make a blockchain. The design of the block consists of three divisions shown in Figure 13.3.

13.8.1 Merkle Tree

A binary hash tree is also known as a Merkle tree. A data structure is an arrangement of a large amount of integrated data that is verified and summarized in a good manner. Merkle tree contains cryptography hashes and arranged in the top to bottom means root display at the top of tree and leaves at the bottom. When a new record (block) is created, this record is hashed, and the hashes are paired; this process is continued until a single hash remains. This single hash is called the root of the Merkle tree.

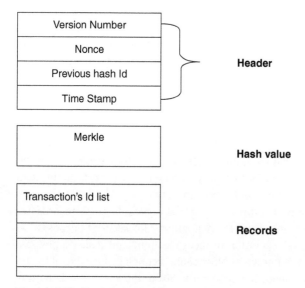

Figure 13.3 Blockchain structure.

13.8.1.1 Advantages of a Merkle Tree

- A Merkle tree provides the validity and integrity of data.
- Less memory and disk space is required in a Merkle tree.
- Merkle tree's computational process is easy and fast.
- A Merkle tree does not require more management for transmitting the information in the network.

13.9 Logical Components of Blockchain

There are four logical components that play a main role in blockchain structure.

Node Applications: Every computer that wants to participate in blockchain must have the specific application related to blockchain and Internet connection. For particular cryptocurrency a computer must run a wallet application related to cryptocurrency. Cryptocurrency allows it to install an application and participate but blockchain has some restrictions.

Shared Ledger: A shared ledger is a part of the distributed ledger managed by node application. After running the blockchain application all nodes are available for the connected users and they utilized the many node applications in a blockchain system.

Consensus Algorithm: One of the most important parts of a node application is to provide the rules of the blockchain system and how it arrives, because every system has different way of arriving for the consensus.

Virtual Machine: Every computer system created a virtual machine by the program that writes instructions using programming language and runs the program according to the capabilities of the virtual machine. When the application is running in a physical computer, the status is indicated within the programming instruction.

13.10 Properties of Blockchain

Immutability is the most important feature of blockchain. The immutability keeps Blockchain permanent. It means no one can be altered or changed [5].

- **Data Protection:** Data and information are stored in the blockchain using the hash value cryptographically. All the information are not stored in their actual form like plain text, but stored with the same encryption algorithm and produces some hash value with the fixed-length, which is unique. The public key is used for transactions, and the private key is used for accessing the data or information.
- **Decentralized:** The network is not controlled by a single authority. It is managed by a group of nodes. Users can directly connect to the Internet and store their assets there, not only cryptocurrency but also digital documents, etc.
- **Distributed Ledger Technology:** There are two ledgers used, public and private; the public ledger is enabled to send all the information related to the transaction

publicly. There is no surety of confidential data. A private ledger is different from the public, but it is unable to hide all the information from the public because the ledger is handled by all users in the network. Distributed ledger technology overcomes the limitation of this system and improves security, and produces a better result.

- **Faster Settlement:** As we know, that traditional finance system transaction process takes time to complete after that customer gets the statement of paper related to that transaction. This issue is now resolved with the invention of blockchain and a customer can process transactions very easily, get a balance or statement within a few minutes,, and its processing fees are nominal.

13.10.1 Ledger

A book that stores all the transaction records of an organization is known as a ledger. After verification of transaction records are stored in the ledger for sharing across the network. Main pool stores the transactions that are not confirmed. All the transactions, either a debit or credit, are arranged in tabular form with separate columns and show the balance amount. Generally, ledgers comprise general, sales, and purchase ledgers.

- **General Ledger:** This type of ledger consists of these types of accounts assets related, expenses related, income-based, or liabilities based.
- **Sales Ledger:** This type of ledger consists of those financial transaction records that the customers do to the financial company.
- **Purchase Ledger:** This type of ledger stores the records related to the spent money by the company for purchasing.

13.11 Types of Blockchain

There are three types of blockchain available private, consortium, and public [6].

- **Private Blockchain:** Private blockchain provides the accessing control only on those people who are authorized by the organization or a group. Actually, private blockchain is created for the organization with some specific aim. The reliabilities depend on the authorized person that is connected to each other with the same outcomes. In a private blockchain architecture, the main protocol decuples from the smart contract. Figure 13.4 shows the private blockchain network.
- **Consortium/Hybrid Blockchain:** Consortium blockchain is available for all people. Everyone can view blockchain information, but if someone wants to add information or connect a node in blockchain, they must have to get permission from those authorized participants. The primary purpose of developing this consortium blockchain for the organization to increased the trust environment among the people, consumers, and society.
- **Public Blockchain:** Public blockchain is open to all participants. They can see all the records that are available in the blockchain. Everyone has the facilities to connect with the Blockchain to add some information or create some new node in blockchain for the transaction, but it is less reliable and efficient. An example of public blockchain is Bitcoin. Figure 13.5 shows the public blockchain network.

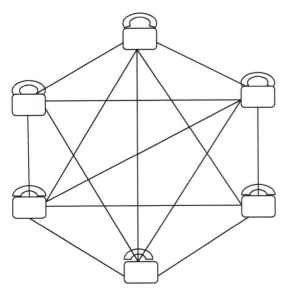

Figure 13.4 Private blockchain network.

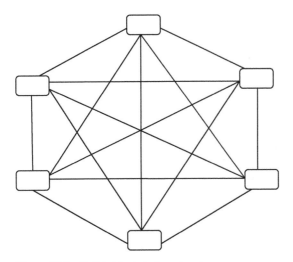

Figure 13.5 Public blockchain network.

13.12   Architecture of Blockchain

The collection of the block that contains the specific information/data is a block-chain. Blocks are genuinely connected and sharing information peer-to-peer like a distributed system instead of centralized system. In blockchain, all the computer devices are linked together and transmitted securely by creating a decentralized network. Centralized, decentralized, and distributed network shows in Figure 13.6 [7].

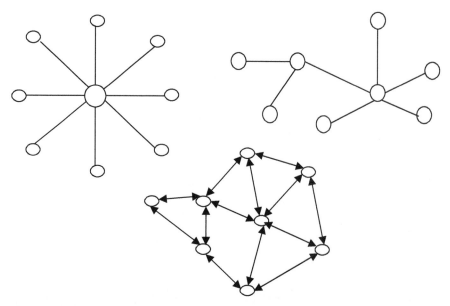

Figure 13.6 Centralized, decentralized, and distributed network. *Source*: Modified from Marco and Lakhani 2017.

Table 13.1 Block identifiers.

Header Hash	Height of Block
A cryptographic hash is a primary identifier.	Height indicates the position of the block.
Block header hash creates a digital fingerprint shown in 32 bytes.	The first block is created with zero height.
The block is defined uniquely with the hash.	Next block shift on the top of the previous block in the chain.
The block hash is separate from the block data.	Block height is not mentioned in block data.

13.13 Block Identifiers

Blockchain is a collection of blocks in a network for sharing information. Header hash and height of block are two identifiers shown in Table 13.1.

13.14 Working of Blockchain

These are the fundamental component of blockchain, such as

- Node: Each system may be user or computer
- Block: Store the transaction details and send them to the network
- Transaction: The building block of blockchain
- Chain: Sequence of block

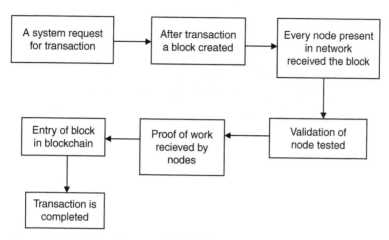

Figure 13.7 Working diagram of blockchain.

The working of blockchain where user wants to join the blockchain system for the transaction of any asset may be tangible or intangible when the data block is created. This data block is a collection of information also known as records, all the data blocks are connected to each other through network and creates a chain for transforming the data from one node to another. After received the node check the validation properly and get the proof of work by nodes and complete the transaction process shown in Figure 13.7.

13.15 Transaction Propagation and Validation in Distributed Network

Blockchain makes the process of a transaction between the nodes simple. After a successful transaction, it propagates in a distributed network. In a distributed network transaction, it is transmitted peer-to-peer taking only 1–2 seconds. It is faster and cost-effective and does not require any mediator between the transactions [8].

Transaction verification is done by the particular node maintained in the ledger and calls the miner, which are he people who add the transaction in blockchain. Miners are individual nodes that hold the copy of the ledger and verifies the transactions happening in the network. Using the cryptographic algorithm, miners validate the trades across the network.

The miners must be able to confirm that:

- The originator of the transaction possesses the funds being transferred.
- The originator of the transaction has obtained the funds by one of the means commonly recognized as valid.
- Once the transaction is verified, it is stored in a shared ledger across the network.
- Unconfirmed transactions will be stored in the main pool area, and from there, the miner will pick and create the block in blockchain for marketing.

13.16   Proof of Work

A piece of data that is costly, time-consuming, and difficult to produce is known as proof of work, but it is easy for others to verify and satisfy certain requirements. Three main components help in achieving the proof of work solutions.

- **Nonce:** A random number whose value is set so that the block's hash will contain a run of leading zeros. The rest of the field may not be changed, as they have a defined meaning.
- **Hash:** A fixed-length number that results in large unchanging data when read.
- **Transaction:** Authentic transfers of bitcoin ownership collected and recorded in the blockchain.

13.17   Blockchain Technologies

Blockchain technology is a structure like a block, which is nothing but storage of records that are linked together and make a chain in a database. This storage is also known as a digital ledger.

13.17.1   Smart Contracts

A simple program of blockchain, which runs when the pre-determined condition meets this program called a smart contract. It is a self-executing agreement between the people or organization [9]. The smart contract is a line of code derived from some rule that writes by the consent of the seller or buyer. Both are agreeing with this. A smart contract is contained across the distributed network and controls the transaction shown in Figure 13.8.

The smart contract creates a trusted environment for the transaction and keeps it safe from central authority or any external enforcement mechanism. Smart contracts provide the facilities where every transaction is trace easily, create transparency, and irreversibility.

An American scientist, Nick Szabo, invented in 1994 a new digital virtual currency, "Bit Gold." A smart contract defines in terms of computer-based transaction protocols that execute all the terms of an agreement. The primary purpose of this smart contract is to develop the functionality of digital transactions.

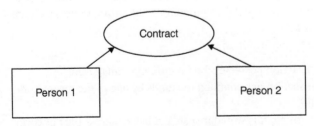

Figure 13.8   Contract between two people.

13.18   Blockchain Applications

Blockchain application is more valuable than cryptocurrency and Bitcoin. It can create transparency, fairness, saving money and time for businesses, and make work more efficient for the government. There are some applications

- Payment processing
- Data sharing
- Digital voting
- Medical recordkeeping
- Managing Internet of Things (IoT) networks
- Real-time IoT operating systems
- Cryptocurrency exchange
- Personal identity security

13.19   Challenges of Blockchain

There are many organizations, policymakers, and business leaders are ready to adopt still some challenges are responsible for slowing the acceptance of this technology such as:

Scalability: The main issue to adopt this technique is scalability. However, thousands of transactions are possible in the network per second without any error. Bitcoin and Etherum show slow transaction processing while ethereum offers the scaling solution that makes blockchain viable.

Interoperability: It is a second major issue that identified the reasons for not adopting this technology. Most of the blockchain does not communicate to another blockchain, meaning they do not share the information. To overcome this problem by developing the smart bridge and fill the gap of communication.

Energy Consumption: Blockchain technique based on the proof-of-work method for valid transactions this method used a lot of power to solve the complex mathematic process. To overcome this problem, the co-founder of Ethereum suggest switching plans from proof-of-work to proof-of-stake. It does not require high energy consumption.

13.20   Conclusion

Cryptocurrency indicates the digital asset design as a medium of exchange currency to secure the record and supports decentralization control. Digital money that is widely used is known as Bitcoin, Ripple, and Etherum. Blockchain application is more valuable than cryptocurrency and Bitcoin. It is a wellstructured technology that stored the information as a record known as a block. Blockchain is based on cryptographic, peer-to-peer network. It believed on decentralization of central system for providing the facilities, all the nodes perform their work individually and complete the task with the coordination of each other and get the outcomes, three types of blockchain available private, consortium, and public. Smart contract is a simple program of blockchain, which is run under the pre-determined condition. It is most popular and easy to use for everyone.

Bibliography

1 Barde, S. and Tikariha, N. (2020). Cyber crime problems and prevention effect on society. *Journal of Information and Computational Science* 13 (1): 29–36.

2 Sagona-Stophel, K. Bitcoin 101 white paper (PDF). Archived from the original (PDF) on August 13, 2016. (Retrieved July 11, 2016).

3 Bitcoin developer chats about regulation, open source, and the elusive Satoshi Nakamoto. Archived October 3, 2014 at the Wayback Machine. PCWorld (May 26, 2013).

4 Llison, I. (September 8, 2015). If banks want benefits of blockchains, they must go permissionless. *International Business Times*. Archived from the original on 12 September 2015, (Retrieved September 15, 2015).

5 Hajric, V. (October 21, 2020). Bitcoin surges to highest since july 2019 after PayPal Embrace. *Bloomberg Law*, (Retrieved October 25, 2020).

6 Melanie, S. (2015). *Blockchain: Blueprint for a New Economy*. O'Reilly Media, Inc.

7 Marco, J. and Lakhani, K. (2017). The truth about blockchain. *Harvard Business Review* 95 (1): 118–127.

8 Michael, C. et al. (2016). Blockchain technology: beyond bitcoin. *Applied Innovation* 2: 6–19.

9 Barde S. (2021). Blockchain-Based Cyber Security. *Transforming Cybersecurity Solutions Using Blockchain, Springer Nature* 55: 2021.

14

Importance of Cybersecurity and Its Subdomains

Parag H. Rughani

Associate Professor, National Forensic Sciences University, India

14.1   Introduction

With the advancement in various technologies and inventions, the number of users in the cyberspace has increased exponentially in the last decade. Availability of high-speed Internet with large data storage and smartphones have changed the life-style of the people in recent time. Smartphones and the Internet have become the most essential living things for human beings in the last few years. The current generation has witnessed migration of mankind from physical to virtual world and this transformation has taken place so rapidly that people couldn't even realize when they became addicted to these high-end technologies. Being the most crowded and virtual space, cyberspace has also attracted criminals toward it. The number of crimes committed in recent years have proved that if a space is not controlled by any jurisdiction then it can become highly vulnerable for legitimate and innocent people.

Looking at the high number of crimes committed and reported in cyberspace, a need to secure cyberspace has been realized by the individuals, organizations, and governments of different regions and countries. The "Cybersecurity Profession" like the terms "cyberspace" and "cybercrimes" has also become one of the most popular terms in the last decade. All these terms are still buzzwords as the use of cyberspace is increasing every day and every minute.

As per the Cybersecurity Ventures' prediction there will be 3.5 million unfilled cybersecurity jobs globally by 2021 [1]. However, it is almost impossible to secure cyberspace by keeping pace with the way in which new technologies are being adopted. The demand to automate and ease work through computers and the Internet is so high that most of the manufacturers and programmers are ignoring the security component in their software/hardware products to meet these requirements. This situation leads to the issues where the cybersecurity experts (as being very low in number) fails in keeping pace with the advanced methods of cybercrimes on highly changing technologies. It has been found that the pace with which the cybercrimes are committed using and targeting emerging technologies is much higher than that by which cybersecurity experts get trained with existing technologies.

Capacity building in the field of cybersecurity and maintaining pace with the emerging technologies is one of the major challenges faced by governments and

Emerging Computing Paradigms: Principles, Advances and Applications, First Edition.
Edited by Umang Singh, San Murugesan and Ashish Seth.

organizations across the world. One of the solutions to handle this situation could be to identify different branches of cybersecurity and train students in specialized areas from their school level. For example, a student studying in the 7th or 8th grade is highly acquainted with the use of computer and Internet technology and can also write programs. Like other domains, if we can identify possible specialization pertaining to different branches of cybersecurity then such students can be trained in those subdomains. Over the period of 5 to 10 years from now, we may have a huge number of focused cybersecurity experts to combat the fight against the cyber criminals and to make the cyberspace highly reliable and secure. This would also help in meeting requirement of cybersecurity experts which would be caused due to the market growth predicted for the next 5 years [2]. A regional chart prepared by Mordor Intelligence for the same is shown in Figure 14.1 below.

14.2 Introduction to Cybersecurity

The term *cybersecurity* is probably becoming one of the most popular terms amongst those who want to make their career in the IT field. Though, almost everyone talks about cybersecurity, the majority have a very vague understanding of the term and its use. In this section, we will try to understand what exactly is cybersecurity and what are its applications?

Before we define the term cybersecurity we must understand parts of the term, i.e., cyber + security. Let us start with the term *security*. As defined by Oxford languages, security means "the state of being free from danger or threat." As being a general word "security" is used more in context with physical security when used alone. In general, we use the security word when something needs to be secured or protected from possible harm or threat. For example, when we say security of a house means ensuring that the house remains secure from all possible threats including natural (like earthquake, flood, etc.) and manmade (theft, fire, etc.).

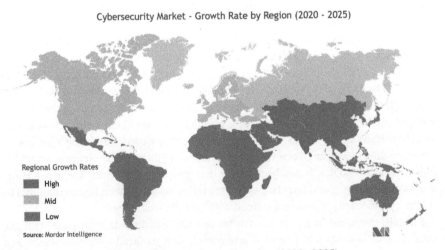

Figure 14.1 Cybersecurity market—Growth rate by region (2020–2025).

Similarly, if we try to understand the term *cyber* then it refers to an ecosystem commonly known as cyberspace that consists of interconnected electronic devices. As these interconnected devices communicate virtually, the cyberspace is also treated as a virtual entity. The short form "cyber" instead of "cyberspace" is used by most of the users.

Now, if we combine these two terms *cyber* and *security*, we can easily say that it refers to security of cyberspace. Most of the users are aware about this aspect but, as we have said earlier cyberspace is a virtual ecosystem and nothing is tangible there, then what exactly needs to be secured? We will need to dive into more detail of cyberspace to get an answer to the above question.

As defined earlier, cyberspace is an ecosystem of interconnected electronic devices. These interconnected devices range from a sensor of an Internet of Things (IoT) infrastructure to a supercomputer. The networks that connect these devices allow them to communicate and transmit information. Using the cyberspace, a computer in the cyberspace can send data in the form of email, picture, video, PDF or even as a chat message to another computer or smartphone as the sender and receiver are interconnected.

The most important and crucial component of a cyberspace is data. The data that is stored in the end points and the data that gets transmitted over the Internet is the wealth of the cyberspace because it may contain someone's confidential information, someone's personal data, someone's official data and a lot.

Securing this data in the cyberspace from the criminals and bad actors is the real cybersecurity. In broad meaning, the data comprise information and hence, we call it information security. Though, information security and cybersecurity terms are used interchangeably. However, many authors claim that information security pertains to the virtual and physical world and cannot be restricted to only cyberspace as physical documents also contain information and their security is considered as physical security rather than cybersecurity. While keeping in mind that data or information is the most precious thing to secure in the cyberspace, we can recall the information security triad, commonly known as CIA triad as shown in Figure 14.2

The confidentiality in this triad refers to the confidentiality of the information or data at rest and in motion. When we refer to confidentiality we want to ensure that the data stored on electronic devices or data being transmitted over the net remain confidential to the intended parties only and no other party can see it. For example, if user A is sending an email or a file to user B then data being transmitted between A and B should not be accessible by any other parties. Similarly, data stored by user

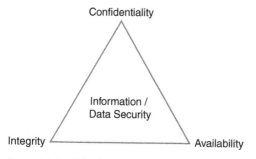

Figure 14.2 CIA triad.

A on his computer should remain confidential to him only and no one else should be able to access and use that data. Cryptography is one of the most frequently used methods to ensure confidentiality.

The integrity on the other hand refers to integrity of the information or data at rest and in motion. Here, integrity can be ensured by making sure that the data remains intake and unchanged. For example, if user A is sending a message to user B then the data being sent should remain intake and user B should receive the exact message as being sent by user A. No one in between can alter the message or can destroy the message. Similarly, data stored by user A on his computer should remain intake and no one should be able to alter the data so that when he accesses it again he can get the original data without any alteration. There are various methods to achieve integrity, hashing and digital signature are the most common ways to ensure it.

Finally, availability, the last component of the CIA triangle refers to availability of data or information. The objective here is to make sure that the data is available to the intended and legitimate users whenever they need to access it. For example, a mailbox (email service provided by a company) of user A should be accessible by him at any point of time. This may not be available in the cases of denial of service or distributed denial of service attacks, which prevent legitimate users from accessing their information. Similarly, data stored by user A on his computer should be available when he wants to access it, again this can be prevented by a ransomware attack which prevents the legitimate user or owner of the data from accessing his data. A combination of various methods is required to be used to ensure availability of information to the legitimate users.

If an organization ensures all three: i.e., confidentiality, integrity, and availability of CIA triad then they can claim that they are able to secure their information in cyberspace. However, it may seem easy and feasible to satisfy the needs of the CIA triad but practically it has a lot of issues and dependencies. The major challenge in this is interconnection between components owned by the organization and external devices. One may take care of all the endpoints, network devices and other components belonging to one but they cannot assure the same for the external entities like Internet Service Providers, third-party websites, clients' infrastructure and all other entities which are virtually connected. This interdependency is something which is difficult to avoid and to handle also.

Another challenge in achieving 100% cybersecurity is the human factor. Though the latest technologies provide highly automated tools, they do not eliminate human intervention completely. The technologies are controlled by human beings and they must be. This poses a severe threat from the cybersecurity's point of view. For example, your computer is fully patched with latest security updates and you have installed highly reliable anti-virus for protection, but if you keep a very simple password which can be compromised easily by the hackers to gain elevated access to your computer then the security provided by the tools and technologies become useless. In fact, a lot of successful cyber-attacks in the past were successful due to the mistakes committed by human beings.

At a nutshell, one can remember that the main objective of cybersecurity or securing cyberspace is to secure data stored and transmitted in cyberspace. The following section discusses few important subdomains of cybersecurity.

14.3   Subdomains of Cybersecurity

In this section we will try to understand different subdomains of cybersecurity. The cybersecurity is like an ocean and explaining it comprehensively is extremely difficult. One can attempt to touch major branches or subdomains of this vast domain but covering all the possible domains is a herculean task. The major reason behind this is the obvious nature of cyberspace. Cyberspace is a highly volatile and ever emerging place. If someone tries to put everything in the form of a book or a chapter to cover subdomains of the cybersecurity then also it will be incomplete as there could be a few completely new technologies that get discovered by the time the content gets published. A relevant example to understand the ever-changing cyberspace is to understand how new types of Artificial Intelligence (AI) based applications are being realized by different domain experts almost every day. The best way to understand these subdomains of cybersecurity in better way is to classify them into broad categories.

In this section, based on my experience and understanding I have tried to classify subdomains of cybersecurity in four major categories namely: 1) target- based cybersecurity domains (what to secure?); 2) tool/methodology based cybersecurity domains (how to secure?); 3) incident response; and 4) emerging and futuristic subdomains of cybersecurity. The aim behind categorizing the subdomains of cybersecurity is to make sure that most of the existing subdomains are covered. Further, this will also help the readers in segregating the subdomains based on their expertise and interest.

14.3.1   Target-Based Cybersecurity Domains (What to Secure?)

The first step to understand cybersecurity is to start with the targets that need to be secured. This section focuses on different entities of cyberspace that could be vulnerable to cyber attacks. Until one knows what he has to secure he cannot decide the methodology or technique that can be adopted for such security. We must remember that the broad objective of cybersecurity is to secure the information or data stored or transmitted in different formats ranging from user files to the instruction stored in sensors of an IoT setup. However, data or information can be stored, retrieved, and interpreted in different ways on different platforms by different applications and that is where we can see the scope of applying cybersecurity in different ways for different situations. This section mainly discusses the security of operating systems, network, applications, hardware, and others.

14.3.1.1   Securing Operating Systems

The main interface between the user and hardware is an operating system. We are aware about most popular computer operating systems like Windows, Linux, and MAC, similarly we are familiar with mobile/smartphone operating systems like Android and iOS. Apart from computers and smartphones, there are dedicated operating systems for different hardware, devices, vehicles, appliances, and electronic gadgets. These operating systems play a crucial role in the way data is being stored and retrieved from the computer or smartphone or any other electronic device. In fact, the operating system is also responsible for sending and receiving data from and to the device. As mentioned earlier, cybersecurity deals with anything where data is present and the operating system is the most important thing from the cybersecurity's perspective. When we talk about different operating systems, we need to correlate security of those operating systems separately as the way Windows works and manages data is very different from the way it is being done in the Linux or MAC OS.

The operating system has always been the first target for the criminals and bad actors as an OS can allow the criminal to gain access to everything on that machine. There are specially designed malware or malicious exploits for different operating systems and even for different versions of different operating systems. This is possible to do as each version or flavor has possible vulnerability that can be exploited differently. In this subdomain of what to secure, I would like to emphasize on OS security and precisely saying security of each possible operating system. Now, based on this discussion you can confidently say that Windows Security, Linux Security, MAC Security, Android Security, iOS Security, and Security of other OSes are branches of cybersecurity.

So, when we say Windows Security, we are referring to a subdomain of cybersecurity that deals with the security of the devices that are running on Windows operating system. At first instance this may seem a very small subdomain of cybersecurity but once we start digging, we can realize that this alone could be taught in a two-year full-time post-graduation or master degree course. The windows security refers to all the associated security aspects of Windows OS and thus one needs to learn and understand how Windows work to ensure security of it. For example, one of the components of Windows machines is the way it handles Dynamic Link Libraries (DLLs). The DLL injection [3] method allows criminals to inject malicious code inside a DLL file, which can lead to severe damage. To prevent this, one should have sufficient knowledge of how DLLs work, how they get loaded inside RAM and how they can be compromised. Similarly, there are a lot of other areas that need to be addressed when we talk about Windows security and thus it makes itself a huge subdomain of cybersecurity and hence, all the operating systems individually do.

14.3.1.2 Securing Applications

The most common way through which intruders enter into a system is by exploiting the vulnerabilities residing in a user level application. When we say applications, we are referring to all sorts of applications ranging from mobile applications to enterprise level applications. All these applications become a gateway for criminals if they are vulnerable. For example, a vulnerability in a PDF reader or an internet browser may allow bad actors to exploit that vulnerability and gain access to the device on which such vulnerable application is running. As mentioned earlier, these applications could be computer applications, mobile applications, web applications, databases, or even system level applications pose threat if they are misconfigured or have any vulnerabilities. Again, each of these applications opens a new branch of study for their security. For example, Android application security itself is a huge domain as it deals with the way Android applications work and handle the information, similarly web application security is another subdomain that deals with all sorts of web applications and their security.

The knowledge of how these applications work and manage information can make you a security expert of that group of applications as you can now think what could go wrong and if anything goes wrong how it can be handled?

14.3.1.3 Securing Hardware

Another component of cyberspace that needs to be secured is the actual hardware. Though a user or even a bad actor can access the hardware through the operating system in most of the cases, there are chances when the hardware itself has vulnerabilities that can be exploited by bypassing the operating systems. The very recent examples of Meltdown [4] and Spectre [5] reveal how dangerous it could be to have some vulnerability on a processor. This branch of hardware security refers to possible ways by which any hardware needs to be secured. Again, the fundamental rule for security is applicable to all targets, i.e., having thorough knowledge of the component that needs to be secured. The hardware or device security can be achieved only when one knows how it works and processes data.

14.3.1.4 Securing Network

Probably, network security is the most popular amongst the security professionals. This branch deals with security of all sorts of networks and can be applied to both: wired and wireless networks. The network security itself is a huge domain as it deals with the data in transit, all the components and media through which data gets transmitted. Security of network devices, protocols, media, and all other associating components fall under this category. Apart from above, the configuration of devices like firewall, Intrusion Detection/Prevention Systems, Log Aggregators are also part of the network security branch of cybersecurity.

14.3.1.5 Other Targets

Though majority of targets are broadly covered in above four subcategories, there are some familiar terms related to cybersecurity, which are heard every now and then. It would be injustice to not mention them here, however, I will try to justify that they are already part of one of the four subcategories mentioned above.

One such popular term these days is cloud security, which refers to the security of the cloud. If we try to understand what is cloud then we can easily map it with operating systems, network and application security or even additionally with hardware security as a cloud compromises of hardware, operating system, and applications. Further, one needs to use a network to access a resource on the cloud. Though it can be put in above categories it can still be considered as a distinct branch as it deals exclusively with clouds and hence has predefined scope. This logic applies to all such categories.

Another popular term is Internet of Things Security or IoT Security, again if you check it from the above categories' perspective then it is nothing more than a combination of hardware, operating system, network, and application security. Again, like cloud security this can also be treated as a separate branch of cybersecurity due to its unique applications.

Mobile security or computer security or server security are other terms that are frequently used and discussed by the security experts and professionals. Now, after having detailed discussion on major subcategories, one can easily map them with one or more of the main four subcategories, i.e., operating system, network, application, and hardware, mentioned above.

The last target that I would like to discuss in this section is critical infrastructure information security or simply, critical infrastructure security. After the invention of latest technologies, the critical infrastructure of a nation like Power Grid, Railway/ Metro Network, Nuclear Plants, Water Dams, Communication networks, ports, and others are now controlled and managed by electronic devices. Being the most important wealth of a nation they are the first target for the state sponsored attacks. One of the most popular case studies of such cyber attacks is of Stuxnet [6]. Though, they are unique in terms of functions and value they carry, at the end they consist of operating systems, network, hardware, and applications and an expert having detailed knowledge of these can help in securing such critical infrastructures. However, I would emphasize to keep it as a separate subcategory due to their value and nature. They are also referred to with the terms like Supervisory Control and Data Acquisition (SCADA) Security and Industrial Control Systems (ICS) security.

In summary, we can say that any hardware or software individually or as a combination with a network that deals with data can be a target for cyber attack and needs to be secured. Since each of them have different ways to store and handle data, security of each of them leads to a branch of cybersecurity.

14.3.2 Tool/Methodology Based Cybersecurity Domains (How to Secure?)

Knowing what to secure is the first step in the security, but that only does not solve the problem. The most important component in cybersecurity is the approach or method required to be used for securing cyberspace. It has been observed in recent years that organizations and individuals have failed in securing their cyber components even after knowing the threats imposed to those devices. One of the major reasons seen in all those cases was related to selection of wrong or inefficient methods to ensure the security. The most common example in-terms of computers is related to outdated software or operating systems. Majority of cyber attackers or criminals exploit old and unpatched operating systems or software. Wannacry [7], the ransomware that shook the world, targeted Windows XP machines. Unfortunately, Windows XP was used by many individuals and organizations even after Microsoft discontinued its support. Surprisingly, Microsoft was able to publish a patch for the vulnerability exploited by WannaCry even before WannaCry had done the damage. This is a clear indication of ignorance toward cybersecurity as the majority of the Windows XP users did not update their machines to patch that vulnerability and that resulted in loss of millions or even billions of USDs. A proactive approach that can help in identifying possible threats in advance and patching them in a timely manner can prevent the majority of cyber attacks and crimes. This section focuses on a few methodologies and approaches that can be considered as a subdomain of cybersecurity. We will try to understand the importance of compliances, assessment and testing, threat intelligence, and other related topics to understand how an organization can equip itself against the war against cybercrimes.

14.3.2.1 Compliances and Auditing

The biggest myth about cybersecurity that prevails among the individuals and organizations is thinking that I or my organization cannot be attacked. I remember one of the cases that involved a small-scale industry in a small town that was a victim of cyber attack. They must have thought that they do not need to worry as no one would be interested in hacking their cyber components. Unfortunately, it was not an IT company but still was hacked due to a misconfiguration in an Internet connected PBX. The criminals do not check the person or organization neither they check the region, they only check for the open vulnerabilities in any device connected to the cyberspace.

The first step of cybersecurity starts by accepting that you or your organization can be attacked and you must plan to prevent or at least handle such attacks. Proper planning for incident response leads to implementation of guidelines and policies. However, once the planning is done the next step is to enforce them with help of regular auditing. This branch of cybersecurity deals with compliances and auditing and has a wide scope as security of an organization heavily depends on the way policies are complied and the rigorous audits that take place. Many big consultants exclusively work on this aspect and have professionals who are expert in compliance and auditing. It would be worth to note that when the organization does not have provision to design or define policies and guidelines they can use existing frameworks like ISO/IEC—27000, COBIT, HIPAA, and others.

14.3.2.2 Vulnerability Assessment and Penetration Testing (VAPT)

While compliance and auditing mostly deal with paperwork and theoretical aspects, VAPT—the crucial branch of cybersecurity deals with technical aspects. It is important to note that most of the cyber attacks are dependent on some sort of vulnerability that resides in the target. Identifying those vulnerabilities in advance can prevent a possible cyber attack pertaining to it.

The Vulnerability Assessment part of VAPT refers to assessment of IT infrastructure of an organization to check and identify possible vulnerabilities. Basically, there are two types of vulnerabilities 1) known vulnerabilities and 2) unknown vulnerabilities. While known vulnerabilities are known publicly, the unknown vulnerabilities are the one that are not discovered and are hidden. In general, the vulnerabilities can include inefficient coding, improper use of memory, misconfigurations, weak encryption or passwords, and more. You must have heard about buffer overflow, that is a vulnerability in which inefficient coding leads to overflow of buffer during the execution. Another common example could be of open ports, which is a type of misconfiguration which allows intruders to gain access to the system through the open ports. One can easily learn more about different known vulnerabilities from the existing databases like National Vulnerability Database (NVD) [8] or CVE MITRE [9]. Apart from NVD and CVE there are other data sources that publish latest vulnerabilities and maintain the database of existing vulnerabilities. It is easy to refer to these databases and understand the known vulnerabilities and also to find a patch for the same. The only concern with respect to vulnerabilities is of unknown vulnerabilities, as they are not known by anyone, not even by developer or designer of that software or hardware respectively. These unknown vulnerabilities remain hidden until they are discovered by someone, if the criminals discover them before the researchers then they can exploit them till a proper patch is not found.

The other part of VAPT refers to penetration testing, where experts try to monitor behavior of the system or network for possible penetrations. This of course has relevance with vulnerabilities as you can penetrate a system or a network when you have exploits available for such vulnerabilities. Though, you do not have much information available for unknown vulnerabilities, you still have a better chance to protect your organization against all the attacks that can be caused due to known vulnerabilities.

There is a myth about VAPT, as many authors restrict the scope of cybersecurity to VAPT. I have heard many people saying that cybersecurity is all about VAPT and nothing else. Though technically VAPT puts you in better condition in handling cyber attacks, there are a lot of other things that play a major role in cybersecurity.

14.3.2.3 Malware Analysis and Research

This branch of cybersecurity is focused on analysis and research of different types of Malicious Software (Malware). Now, if we put things in a particular sequence then we can correlate malware with the vulnerability. A vulnerability in a system can be compromised with help of an exploit written for it. These exploits are nothing but scripts or code snippets that are written to target the known vulnerability. These exploits are created with malicious intentions and hence fall in the category of malware. For example, a backdoor written to exploit a vulnerability is a malware. Though vulnerabilities and malware are related, there may not be a direct visible relation between them. For example, a ransomware may not exploit any vulnerabilities directly but indirectly it can infect machines successfully only if the system does not have antivirus installed in it and that is nothing but a vulnerability caused by the user. The user left the system vulnerable by not keeping anti-virus, which could prevent ransomware infection.

Malware analysis and research are two different approaches to handle malicious software and both of them are crucial from cybersecurity's point of view. The analysis part refers to the analysis of existing malware to understand their behavior and indicators while research refers to detailed study of unknown malware. Again, similar to applications and operating systems, the malware also falls into different categories like Windows Malware, Linux Malware, Android Malware, IoT Malware, and so on. These malwares can be written for different purposes and hence they also create separate subclasses or families like Trojan, Backdoor, Ransomware, Warm, and others. The researchers also use honeypots or honeytraps to attract new malware. The analysis in this field is mostly done either in a simulated environment or through reverse engineering.

Since malware is at the base of the majority of cyber attacks, this branch of cybersecurity has huge potential. The knowledge of executables in addition to knowledge of operating systems and memory management would be required to get expertise in this field.

14.3.2.4 Threat Intelligence and Log Monitoring

One of the emerging branches which help in efficient cybersecurity is threat intelligence. This branch lets cybersecurity experts gather intelligence about latest threats and happening within the organization and from all over the world. The logs generated by the systems and devices of the organization are collected, parsed, and monitored with help of automated tools like Security Incident and Event Management (SIEM) tools, which give an insight about the activities of the organization and can help in tracking and tracing the malicious events in advance. On the other hand, intelligence about latest threats, incidents, and vulnerabilities from all over the world can help the organization in understanding possible threats and to prepare for them. The Security Operation Center (SOC) is a dedicated center involved in threat intelligence and handling security operations. Many times, SOCs are used in place of Computer Emergency Response Teams (CERTs) or Computer Security Incident Response Team (CSIRTs), however SOCs and CERTs or CSIRTs have different role to play, still their functions can be combined in a single unit depending on companies' requirements. The CERT and CSIRT are discussed in the next subdomain called incident response.

In summary, we can say that appropriate methods and approach can help in implementing better cybersecurity for an organization. However, in this section we have discussed only major branches of cybersecurity from the methodology's point of view, there could be many other ways by which an organization can handle the cyber attack effectively but it is beyond the scope of this chapter to cover all of them.

14.3.3 Incident Response

Responding to an incident is the key strength of any organization. Many times, it does not matter how well prepared you are? The only thing that matters is how well you defend an attack. There are chances that an organization becomes a victim of a cyber attack and in fact it is very difficult to prevent all the cyber attacks. This is applicable in all the sectors in the world. The most recent example of COVID-19 is very much relevant to this section. There is merely any difference between the biological virus and a computer virus. Hence, it is very obvious and bound to happen that even having all sorts of precautions, your cyberspace becomes victim of any cyber attack. Looking at this very natural course of action, one should also be prepared for all possible ways to handle any unwanted incidents.

Effective incident response can reduce the damage at a larger scale and can also prevent many attacks if the incidents are identified in the advanced stage. Handling the incident with proper strategy and plan is extremely important from the cybersecurity's perspective.

This section talks about various methods and steps involved in effective and efficient incident response. One of the important steps that lead to effective cybersecurity is called lessons learned. The lessons learned from an incident becomes a guideline for the future incidents. As an example, in case of a distributed denial of service attack, a website owner may realize that lack of log monitoring facility caused the attack as they could not interpret the logs related to that attack. This type of lesson from such an incident can become a crucial point to prevent similar attacks in future.

In fact, the security always comes after the investigation of a crime. As shown in Figure 14.3, the base of any cybersecurity against a specific attack or crime is highly dependent on the accurate investigation of similar crimes committed with similar modus operandi.

The red arrow from security to crime indicates that security applied based on the investigation can be used to prevent similar type of crimes in future. For example, investigation of an attack revealed that a malware exploits vulnerability X on a

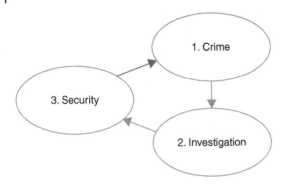

Figure 14.3 Crime, investigation, and security.

Windows 10 machine then the security experts will patch that vulnerability X on Windows 10 machine so in the future the same malware or any other malware similar to that cannot exploit the same vulnerability on the same platform. Hence, the security is always decided based on the past incidents and prevents the attacks committed with similar modus operandi. Someone can argue that if a cybersecurity expert envisions a possible method to commit the crime that does not exist and he applies a patch to that then in such cases security comes even before the crime was committed. This is completely true but the cybersecurity expert who envisions such attacks actually gains the knowledge by thinking the way a criminal will do and the security is derived from that is a result of a crime that had taken place in imagination rather than in real.

The Incident Response Handling and Management is one of the crucial subdomains of cybersecurity, many authors do not consider this as a part of cybersecurity as it majorly deals with investigation and forensics aspects. Since, it comes into picture when the incident is taking place or while the incident has already taken place, it is mostly considered as a post incident activity and thus it is ignored from the cybersecurity's domain. But, as explained above I personally feel that incident response is a part of this domain as it reveals a lot of useful information that can become a base for implementing better cybersecurity. Many times, it has been observed that many organizations can learn from incident handling of incidents taken place at other organizations. For example, if bank A experiences a ransomware attack, they can share the methods used in incident handling and outcomes of the same with other banks that ultimately will help other banks to put necessary measures in preventing similar attacks.

The CERT or the CSIRT are used for efficient incident handling and management by the organizations across the world. The incident response begins with drafting appropriate incident response plan and defining policies to handle any incidents. Following could be stages of an incident handling process (see Figure 14.4).

Discussion of stages of incident response is beyond scope of this chapter, but one of the important steps during the incident response is digital forensics, which deals with forensic analysis of digital evidence. This is one of the branches which itself is like the main cybersecurity domain. When we talk about forensics we can imagine branches belonging to it. A few such branches are computer forensics, mobile forensics, network forensics, cloud forensics, IoT forensics, Windows forensics, Linux

Figure 14.4 Stages of incident response.

forensics, Mac forensics, iOS forensics, Android forensics, and others. The major difference between security and forensics is the time when they are applied. The security is used to prevent the attacks while forensics is used once the crime is committed or incident has taken place. With reference to earlier discussion, forensics being part of investigation, plays a major role in deciding cybersecurity.

In summary, we can say that the incidents are bound to happen and having proper plan and strategies to handle them can help an organization in reducing the possible damage which makes it one of the crucial subdomains of cybersecurity in addition to digital forensics.

14.3.4   Emerging and Futuristic Domains of Cybersecurity

As being the fastest changing domain, the cyberspace and its technologies are highly volatile. What is new today becomes outdated tomorrow, in fact the latest technologies are having very short life. The best example of this is the frequent updates published by different applications. Though, these updates may not always be related to security but they reflect the nature of frequent changes caused due to change in the technology or expectation of the user. It may sound a little strange that how these frequent changes can affect cybersecurity, but both of them are closely related. One of the aspects of this section, when we say emerging domains of cybersecurity, is to discuss emerging threats like adverse use of AI in committing cyber crimes while on the other hand it focuses on use of latest technologies like Drones, AI, IoT, Blockchain and others in combating cybercrimes.

14.3.4.1   Emerging Threats
In this section of futuristic domains of cybersecurity, we will try to explore and understand possible future threats that do not exist at this moment but may become highly important in a decade or two. Many scientists and researchers are able to foresee and discuss some of the future cyber attacks pertaining to the space technologies, self-driving vehicles, robots, critical infrastructure, AI-based solutions, and others. A basic introduction to some of them is given below.

One of the emerging threats is adverse use of AI, deepfake [10] is one such problem. The power of AI can be heavily misused if not used in a controlled environment and this is probably the most dangerous threat not only for cyberspace but for mankind, too. Securing your cyberspace against such AI-based crimes would be highly demanding and will require exceptional skills.

Another such emerging threat is caused by drones, the Unmanned Aerial Vehicles (UAV) or most commonly known as drones can be used by the criminals to commit crimes. The misuse of drones is already experienced in recent years and can lead to more deadly attacks in the future. The drones can also be a powerful weapon in future warfare [11]. Handling drone and drone-based attacks could be a severe challenge in future. The counter measures or anti-drone techniques and drone forensics could be highly important branches of cybersecurity in near future.

Third important emerging area is the use of autonomous vehicles. In future we may hear cases where autonomous vehicles may be involved in killing someone [12]. Making sure that these autonomous vehicles are highly accurate and do not cause any fatality would be the most interesting branch to explore from the cybersecurity's point of view.

A lot of research has been done in the possible threats that could be caused by IoT-based solutions and especially in terms of smart solutions. The smart cities, smart agriculture, smart healthcare, and other such solutions would be controlled by the computers and network and can be used by criminals for malicious intentions [13]. Securing them to make sure they are not exploitable would be a challenge and will require extraordinary skilled manpower in future.

If we see the emerging domains in the cyberspace then we can easily associate security aspects with them as when a new technology is invented it brings threats and challenges with it. The new technologies are aimed more to provide facility and easiness to end users but are less focused toward the security aspects and which make them more vulnerable to cyber attacks.

14.3.4.2 Futuristic Solutions

While the latest technologies like AI, drones, smart technologies, and blockchains can be a good target for the criminals the same technologies can be used in combating crimes. This portion of the chapter discusses possible ways by which some of the emerging technologies can be used in defeating the cyber attacks.

AI can be a game changer in the field of cybersecurity, in fact many AI-based solutions are available at this moment which are highly effective in predicting and preventing cyber attacks [14]. As an example, the SIEM tools, as discussed earlier can help an organization in monitoring logs and events. If these SIEM tools get power of AI then they will become highly effective, in fact some of the SIEM tools like IBM QRadar and others have already started offering machine learning or AI-based features. Similarly, computer vision and natural language processing, the two highly useful branches of machine learning can help security experts in designing solutions to achieve best cybersecurity.

Like AI, blockchain is another emerging technology that can become a powerful branch of cybersecurity [15]. Due to its decentralized characteristics, blockchain can be used in developing robust and reliable solutions to mitigate cyber attacks. Another emerging area that could be very helpful in implementing effective cybersecurity is UAVs or drones. The drones can be used by defense and law enforcement agencies in preventing crimes. As mentioned earlier, a lot of such emerging technologies can equip the existing cybersecurity solutions with better accuracy and performance. Since, it is beyond the scope of this chapter to cover all the latest technologies, only a few important of them are discussed in this subsection.

In summary, I would like to emphasize the use of emerging and futuristic technologies, they can be used for good or bad purposes. The same technology can either be used to commit crime or to prevent crime as we have seen in the case of AI. From cybersecurity's point of view if these technologies are used with bad intentions then there needs to have some mechanisms to prevent them which opens new branches of cybersecurity, on the other hand use of these technologies in creating better cybersecurity solutions also open new branches of cybersecurity. Either way there are a lot of areas where combination of cybersecurity and future technologies can lead to new subdomains of cybersecurity.

14.4 Conclusion

Cybersecurity is probably going to be the most important concerns in the coming years for the governments from all over the world. From developing countries to the developed countries, every government is investing more and more funds behind securing cyberspace and the time is not far when this amount would be almost equivalent or even higher than what is being spent behind physical security. There is no doubt that the cybersecurity profession will be the most demanding profession in near future and experts of this field will be in high demand. This chapter will help all the aspirants who want to make their career in the field of cybersecurity in understanding fundamentals of cybersecurity and various subdomains of cybersecurity and acquiring necessary expertise in that domain.

Bibliography

1 Morgan, S. (2021). Top 5 cybersecurity facts, figures, predictions, and statistics for 2021 to 2025. *Cyber Crime Magazine* (January).
2 Cybersecurity Market – Growth, Trends, COVID-19 Impact, and Forecasts (2021–2026) published by Mordor Intelligence.
3 Javaheri, D. and Hosseinzadeh, M. (2020). A solution for early detection and negation of code and DLL injection attacks of malwares. *Journal of Advanced Defense Science and Technology* 10 (4): 393–406.
4 Lipp, M., Schwarz, M., Gruss, D., Prescher, T., Haas, W., Fogh, A., Horn, J., Mangard, S., Kocher, P., Genkin, D., and Yarom, Y. (2018). Meltdown: reading kernel memory from user space. *27th Security Symposium (Security 18)*, 973–990.
5 Kocher, P., Horn, J., Fogh, A., Genkin, D., Gruss, D., Haas, W., Hamburg, M., Lipp, M., Mangard, S., Prescher, T., and Schwarz, M. (2019, May). Spectre attacks: exploiting speculative execution. *2019 IEEE Symposium on Security and Privacy (SP)*, 1–19. IEEE.
6 Trautman, L. J. and Ormerod, P. C. (2017). Industrial cyber vulnerabilities: lessons from Stuxnet and the Internet of Things. *University of Miami Law Review* 72: 761.
7 Clarke, R. and Youngstein, T. (2017). Cyberattack on Britain's National Health Service—a wake-up call for modern medicine. *New England Journal of Medicine* 377 (5): 409–411.
8 Booth, H., Rike, D. and Witte, G. (2013). The National Vulnerability Database (NVD): Overview, ITL Bulletin, National Institute of Standards and Technology, Gaithersburg, MD.

9 Mell, P. and Grance, T. (2002). Use of the common vulnerabilities and exposures (cve) vulnerability naming scheme. National Inst of Standards and Technology Gaithersburg md Computer Security Div.

10 Kwok, A. O. and Koh, S. G. (2020). Deepfake: a social construction of technology perspective. *Current Issues in Tourism* 1–5.

11 Gusterson, H. (2016). *Drone: Remote Control Warfare*. MIT Press.

12 De Sio, F. S. (2017). Killing by autonomous vehicles and the legal doctrine of necessity. *Ethical Theory and Moral Practice* 20 (2): 411–429.

13 Baig, Z. A., Szewczyk, P., Valli, C., Rabadia, P., Hannay, P., Chernyshev, M., Johnstone, M., Kerai, P., Ibrahim, A., Sansurooah, K., and Syed, N. (2017). Future challenges for smart cities: cyber-security and digital forensics. *Digital Investigation* 22: 3–13.

14 Zeadally, S., Adi, E., Baig, Z., and Khan, I.A. (2020). Harnessing artificial intelligence capabilities to improve cybersecurity. *IEEE Access* 8: 23817–23837.

15 Xu, S. (2020). *Blockchain-Based Cybersecurity Management (B2CSM): Design, Analysis, and Prototype Implementation*. University of Texas at San Antonio San Antonio United States.

Index

Emerging Computing Paradigms: Principles, Advances and Applications, First Edition.
Edited by Umang Singh, San Murugesan and Ashish Seth.
© 2022 John Wiley & Sons Ltd. Published 2022 by John Wiley & Sons Ltd.